Neighborhood Decline

T0346599

The global financial and economic crisis that hit the world since 2008 has affected the lives of many people all over the world and resulted in declining incomes, rising unemployment, fore-closures, forced residential moves, and cut-backs in government expenditure. The extent to which the crisis has affected urban neighborhoods and has led to rising intra-urban inequalities, has not yet received much attention. The implemented budget cuts and austerity programs of national and local governments are likely to have hit some neighborhoods more than others. The authors of this book, which come from a variety of countries and disciplines, show that the economic crisis has affected poor neighborhoods more severely than more affluent ones. The tendency of the state to retreat from these neighborhoods has negative consequences for their residents and may even nullify the investments that have been made in many poor neighbor-hoods in the recent past.

This book was originally published as a special issue of *Urban Geography*.

Ronald van Kempen (1958–2016) was a Professor of Urban Geography at the Faculty of Geosciences, Utrecht University, the Netherlands. His research focused on urban spatial segre-gation, urban diversity, housing for low-income groups, urban governance and its effects on neighbourhoods and residents, social exclusion, and minority ethnic groups. He has published over 200 reports and articles, most of them in international social and urban geography journals.

Gideon Bolt is an Assistant Professor of Urban Geography and Methods & Techniques at the Faculty of Geosciences, Utrecht University, the Netherlands. His research focuses on urban policy, social cohesion, residential segregation, and neighbourhood choice. He is project coor-dinator of the EU-FP7 project DIVERCITIES (Governing Urban Diversity).

Maarten van Ham is a Professor of Urban Renewal and Housing at Delft University of Technology, the Netherlands; and a Professor of Geography at the University of St Andrews, UK. Maarten studied economic geography at Utrecht University, where he obtained his PhD in 2002. Maarten has published over 60 academic papers and 6 edited books, and has expertise in the fields of urban poverty and inequality, segregation, residential mobility and housing choice, and urban and neighbourhood change. In 2014, Maarten was awarded a €2 million ERC Consolidator Grant for a 5-year research project on neighbourhood effects (DEPRIVEDHOODS).

Neighborhood Decline

Edited by
**Ronald van Kempen, Gideon Bolt and
Maarten van Ham**

Routledge
Taylor & Francis Group

LONDON AND NEW YORK

First published 2018
by Routledge

2 Park Square, Milton Park, Abingdon, Oxfordshire OX14 4RN
52 Vanderbilt Avenue, New York, NY 10017

Routledge is an imprint of the Taylor & Francis Group, an informa business

First issued in paperback 2019

Chapter 1 © 2018 Taylor & Francis
Chapter 2 © 2018 Merle Zwiers, Gideon Bolt, Maarten van Ham and Ronald van Kempen
Chapters 3–5 © 2018 Taylor & Francis
Chapter 6 © 2018 Roger Andersson and Lina Hedman
Chapter 7 © 2018 Taylor & Francis

British Library Cataloguing in Publication Data
A catalogue record for this book is available from the British Library

ISBN 13: 978-1-138-74470-7 (hbk)
ISBN 13: 978-0-367-22975-7 (pbk)

Typeset in Minion Pro
by RefineCatch Limited, Bungay, Suffolk

Publisher's Note
The publisher accepts responsibility for any inconsistencies that may have
arisen during the conversion of this book from journal articles to book chapters,
namely the possible inclusion of journal terminology.

Disclaimer
Every effort has been made to contact copyright holders for their permission to
reprint material in this book. The publishers would be grateful to hear from any
copyright holder who is not here acknowledged and will undertake to rectify
any errors or omissions in future editions of this book.

Contents

Citation Information

The chapters in this book were originally published in *Urban Geography*, volume 37, issue 5 (2016). When citing this material, please use the original page numbering for each article, as follows:

Chapter 1
Neighborhood decline and the economic crisis: an introduction
Ronald van Kempen, Gideon Bolt and Maarten van Ham
Urban Geography, volume 37, issue 5 (2016), pp. 655–663

Chapter 2
The global financial crisis and neighborhood decline
Merle Zwiers, Gideon Bolt, Maarten van Ham and Ronald van Kempen
Urban Geography, volume 37, issue 5 (2016), pp. 664–684

Chapter 3
Reclaiming neighborhood from the inside out: regionalism, globalization, and critical community development
Kathe Newman and Edward Goetz
Urban Geography, volume 37, issue 5 (2016), pp. 685–699

Chapter 4
The US Great Recession: exploring its association with Black neighborhood rise, decline and recovery
Derek Hyra and Jacob S. Rugh
Urban Geography, volume 37, issue 5 (2016), pp. 700–726

Chapter 5
Neighborhood change beyond clear storylines: what can assemblage and complexity theories contribute to understandings of seemingly paradoxical neighborhood development?
Katrin Grossmann and Annegret Haase
Urban Geography, volume 37, issue 5 (2016), pp. 727–747

Chapter 6
Economic decline and residential segregation: a Swedish study with focus on Malmö
Roger Andersson and Lina Hedman
Urban Geography, volume 37, issue 5 (2016), pp. 748–768

Chapter 7

Are neighbourhoods dynamic or are they slothful? The limited prevalence and extent of change in neighbourhood socio-economic status, and its implications for regeneration policy
Rebecca Tunstall
Urban Geography, volume 37, issue 5 (2016), pp. 769–784

For any permission-related enquiries please visit:
http://www.tandfonline.com/page/help/permissions

Notes on Contributors

Roger Andersson is a Professor at the Institute for Housing and Urban Research and the Department of Social and Economic Geography, Uppsala University, Sweden. His research is concerned with residential segregation, intra-urban migration, urban policy, and economic and social integration of immigrants.

Gideon Bolt is an Assistant Professor of Urban Geography and Methods & Techniques at the Faculty of Geosciences, Utrecht University, the Netherlands. His research focuses on urban policy, social cohesion, residential segregation, and neighbourhood choice. He is project coordinator of the EU-FP7 project DIVERCITIES (Governing Urban Diversity).

Edward Goetz is a Professor in the Urban and Regional Planning Area at the Humphrey School of Public Affairs, University of Minnesota, Minneapolis, MN, USA. His main research interests are housing policy, urban planning, race and ethnicity, and income inequality and poverty.

Katrin Grossmann is a Professor at the Faculty of Architecture and Urban Planning, University of Applied Sciences Erfurt, Germany.

Annegret Haase is a Scientific Collaborator at the Department of Urban and Environmental Sociology, Heimholtz-Centre for Environment Research, UFZ, Leipzig, Germany. Her expertise lies in international comparative urban research and the social modelling of urban processes.

Lina Hedman is a Research Fellow at the Institute for Housing and Urban Research and Senior Lecturer at the Department of Social and Economic Geography, Uppsala University, Sweden.

Derek Hyra is an Associate Professor at the Department of Public Administration and Policy, American University, Washington DC, USA. His research focuses on processes of neighbourhood change, with an emphasis on housing, metropolitan politics, and race.

Kathe Newman is an Associate Professor at the Edward J. Bloustein School of Planning and Public Policy and Director of the Ralph W. Voorhees Center for Civic Engagement, Rutgers University, Brunswick, NJ, USA. Her research has explored gentrification, foreclosure, urban redevelopment, and community participation.

Jacob S. Rugh is an Assistant Professor at the Department of Sociology, Brigham Young University, Provo, UT, USA. His research interests are race and space, housing segregation, mortgage lending, and stratification.

Rebecca Tunstall is the Director of the Centre for Housing Policy, University of York, UK. Her principal areas of work have been social housing, neighbourhoods, and inequality.

Maarten van Ham is a Professor of Urban Renewal and Housing at the Faculty of Architecture and the Built Environment, Delft University of Technology, the Netherlands; and a Professor of Geography at the School of Geography and Geosciences, University of St Andrews, UK. He has published over 60 academic papers and 6 edited books, and has expertise in the fields of urban poverty and inequality, segregation, residential mobility and housing choice, and urban and neighbourhood change. In 2014, Maarten was awarded a €2 million ERC Consolidator Grant for a 5-year research project on neighbourhood effects (DEPRIVEDHOODS).

Ronald van Kempen (1958–2016) was a Professor of Urban Geography at the Faculty of Geosciences, Utrecht University, the Netherlands. His research focused on urban spatial segregation, urban diversity, housing for low-income groups, urban governance and its effects on neighbourhoods and residents, social exclusion, and minority ethnic groups. He has published over 200 reports and articles, most of them in international social and urban geography journals.

Merle Zwiers is a PhD student at the Faculty of Architecture and the Built Environment, Delft University of Technology, the Netherlands. Her research interests are neighborhood effects, inequality, urban and neighborhood change, and innovative methodologies.

This book is dedicated to the memory of
Ronald van Kempen
1958–2016

Neighborhood decline and the economic crisis: an introduction

Ronald van Kempen[a], Gideon Bolt[a] and Maarten van Ham[b,c]

[a]Faculty of Geosciences, Utrecht University, Utrecht, The Netherlands; [b]OTB - Research for the Built Environment, Faculty of Architecture and the Built Environment, Delft University of Technology, Delft, The Netherlands; [c]Centre for Housing Research, School of Geography and Geosciences, University of St Andrews, St Andrews, Fife, Scotland

ABSTRACT
Urban neighborhoods are still important in the lives of its residents. Therefore, it is important to find out how the recent global financial and economic crisis affects these neighborhoods. Which types of neighborhoods and which residents suffer more than others? This introduction provides an overview of the papers in this special feature that focus on this question. It concludes with the statement that governments should specifically pay attention to the poor neighborhoods and the people living there, because here the effects of the crisis are very prominent and in many cases probably long-lasting.

Background

Since the Chicago School in the 1920s (e.g., Hoyt, 1939; Park, Burgess, & McKenzie, 1925/1974), neighborhoods have never been off the agenda of urban researchers. This is evidenced by the numerous articles and books that focus on life in urban neighborhoods (e.g., Gans, 1962; Suttles, 1974), on urban spatial segregation (e.g., Bolt, Van Kempen, & Van Ham, 2008; Kazepov, 2005; Lieberson, 1981; Logan, Stults, & Farley, 2004; Musterd & Ostendorf, 1998; Musterd & Van Kempen, 2009; South, Crowder, & Chavez, 2005; Taeuber & Taeuber, 1965), and the somewhat more recent research on neighborhood effects (Friedrichs, Galster, & Musterd, 2003; Musterd, Andersson, Galster, & Kauppinen, 2008; Van Ham & Manley, 2010). Many studies show that the neighborhood is certainly of importance, in particular for poorer households (see, e.g., Ellen & Turner, 1997; Guest & Wierzbicki, 1999) and for specific ethnic groups in the expectation that they will be more likely to receive social, economic, and emotional support from their fellow residents (Enchautegui, 1997; Fong & Gulia, 1999).

However, since the 1960s, researchers have also suggested that the neighborhood is becoming less important in people's lives (Stein, 1972; Webber, 1964). At that time, this idea was based on the rapidly increasing (auto)mobility of urban society. Later, under the influence of the globalization literature, the further internationalization of the economy

(Castells, 2000; Marcuse & Van Kempen, 2000), and the rapid development of the Internet and social media, the idea that the local becomes less important than the global became the dominant discourse (Graham & Marvin, 1996; Van Kempen & Wissink, 2014; Wellman, 1999, 2001). In times of globalization, better transport and the Internet, many urban residents can "go" almost anywhere, so why bother about the neighborhood?

At the same time, there is strong evidence that neighborhood remains important in the lives of people: we are willing to pay considerably more for a house in a good neighborhood (Cheshire, 2012); children go to local schools, play in the street, and have a lot of their friends and social contacts in the neighborhood; local contacts and networks are still important in people's lives; and local residents organize themselves when there are threats to their neighborhood. In many neighborhoods, social activities are organized by neighbors, often aimed at enhancing social contacts within the neighborhood. On the other side of the spectrum, we also know that problems in the local environment can negatively affect peoples' lives and lead to tensions between residents. In such cases, life is far less agreeable, even if people have their leisure and work activities in other parts of the city. Such problems may even be a reason to move to another neighborhood as quickly as possible (e.g., Clark & Ledwith, 2006). So, neighborhoods still matter for many.

A neighborhood can be defined as a relatively small spatial subdivision of a city or town for which a number of physical and socioeconomic characteristics can be measured (for a more elaborate discussion of this definition, see Zwiers, Bolt, Van Ham, & Van Kempen, 2016, this issue). Cities always comprise a mosaic of neighborhoods: poor neighborhoods, slums, ghettos (not everywhere), mixed neighborhoods, "white" neighborhoods, immigrant neighborhoods, posh neighborhoods, working-class neighborhoods, deprived neighborhoods, gentrified areas, gated communities (also not everywhere), calm suburbs, and bustling inner cities. Some neighborhoods show great stability over time. Others gradually, or sometimes quite quickly, change for better or for worse, e.g., as the consequence of neighborhood-directed policies or because of the availability of new and attractive housing opportunities elsewhere in the city or region (causing people who can afford it to move away). Processes such as gentrification can also rapidly alter the population and the character of neighborhoods.

A large number of papers and books have been dedicated to neighborhood change, and especially neighborhood decline, with contributions from various countries, including some more theoretical accounts of neighborhood decline (e.g. Grigsby, Baratz, Galster, & Maclennan, 1987; Prak & Priemus, 1986; Temkin & Rohe, 1996). The discussion of neighborhood decline combines population change, physical decline, economic development, and the role of governance and policy. We define neighborhood decline as *any negative development in the physical, social or economic conditions of a neighborhood as experienced by its residents or other stakeholders* (see Zwiers et al., 2016, this issue). Neighborhood decline often combines a number of negative developments, such as a declining physical quality of the housing stock, the outflow of more affluent households, the inflow of less affluent households, an unfriendly or even dangerous atmosphere in the streets, rising criminality, etc. The combination of all such developments can easily lead to a spiral of decline. Most models that aim to shed light on the causes of neighborhood decline also pay attention to external factors, especially (global) macro-developments.

The global financial and economic crisis that hit the world since 2008 is an example of a macro-development that affects neighborhoods. The beginning of the crisis was signaled by the fall of Lehman Brothers in September 2008, and widened to encompass banks and other financial institutions, multinationals and local firms, the employment structure, the housing market, property values, and public expenditure. But maybe most importantly, it affected the lives of many people all over the world in the form of declining incomes, unemployment, foreclosures, forced moves, and reduced services as national and local government expenditures were cut back.

The extent to which the crisis also affected urban neighborhoods has not yet received much attention. This is unfortunate, because although it is true that the crisis has affected all countries, regions, and neighborhoods, there are large geographical differences with respect to the impact of the crisis. Some countries have a much larger social rented sector than others, which may cushion the effects of the crisis. Some governments have decided to implement more rigid austerity programs and budget cuts than others. Those budget cuts are likely to hit some regions and neighborhoods more than others. The poorest neighborhoods are likely to be hit most by the crisis, as their population runs the biggest risk of declining incomes and job loss. The effects of the crisis also partially depend on the definition (especially the size) of the neighborhood: in large heterogeneous neighborhoods, effects may be more differentiated than in small homogeneous (e.g., very rich or very poor) ones. For this special feature of *Urban Geography*, we asked a number of international authors to reflect on the development of urban neighborhoods in times of crisis.

Content of the special issue

The first paper in this special issue (Zwiers et al., 2016) attempts to unravel the complex and multidimensional process of neighborhood decline. Although many models of decline have been proposed, and many factors have been identified as a cause of neighborhood decline, until now, researchers have paid only limited attention to the effects of economic developments and of the economic crisis on neighborhoods. The authors of this introductory paper formulate a number of hypotheses that can be seen as starting points for further research into the relationship between general economic developments and developments in and of neighborhoods.

Kathe Newman and Edward Goetz (2015, this issue) argue that the question of whether neighborhoods are still important in a globalizing world is not a crucial one. They point to the increasing interrelatedness of neighborhoods and globalization: what happens in one place may affect what happens on the other side of the world. They, e.g., state that changing financial regulations in the United Kingdom could affect the cost and availability of capital in the United States (and elsewhere), which can profoundly affect investments in urban places. The global economic crisis thus influences local developments (although it is not always clear exactly how this works). Neighborhoods remain at the heart of community development policy and practice, but part of the community development agenda is to understand and engage with external processes that shape the development of neighborhoods.

In their paper, Newman and Goetz also point to a second issue: the growing body of literature and policy action that privileges the region as the place from which to understand urban decline and which addresses issues that have historically been the concern of community development. Newman and Goetz define regionalism as a political and policy approach that locates the problems of central city neighborhoods, as well as the solutions to those problems, in the relationship of those neighborhoods to larger, metropolitan-level economic, social, and political dynamics. This idea is very much related to the fact that research has indicated that place-based revitalization policies have not effectively reversed the decline of central city neighborhoods in (American) cities. Indeed, too much neighborhood-focused work does not engage with the broader economic and political processes that help to produce local conditions. At the same time, Newman and Goetz identify potential danger in the work of the (primitive) regionalists: they often pathologize deprived neighborhoods, and propose solutions that involve moving people away from those areas. This is related to policies in Western Europe that focus on the restructuring of urban neighborhoods (i.e. large-scale demolition in order to break down spatial concentrations of the poor), leading to forced moves of low-income households to other places within urban regions. Such policies often fail to improve the socioeconomic situation of households living in deprived communities.

In their longitudinal multiple case study, Derek Hyra and Jacob Rugh (2016, this issue) compare three gentrifying African American communities: Bronzeville in Chicago, Harlem in New York City, and Shaw/U Street in Washington, DC. Much of each neighborhood's older housing stock was constructed in the mid to late nineteenth century, when these areas were still predominantly affluent and middle class. In the 1980s and 1990s, these neighborhoods were considered as "no go" zones with high levels of poverty and crime, but in the 1990s they started to revitalize. In the 2000s, during the subprime lending boom, property values began to skyrocket (Hyra & Rugh, 2016, this issue). The Black population steadily declined in number. In their paper, the authors compare the development of neighborhoods across three phases: pre-Recession (2000–2006), Recession (2007–2009), and post-Recession (2010–2012) periods.

Hyra and Rugh emphasize the effects of subprime lending. Their main conclusion is that neighborhoods are differentially influenced by the economic crisis and they make clear that distinct community and city contexts, in particular racial and class neighborhood transitions and citywide unemployment and housing market conditions, mediate the influence of national economic decline and recovery. They conclude that elite upper class, mixed-race gentrification in Harlem and Shaw/U Street, compared to Bronzeville's Black middle-class gentrification, might have protected these communities from excessive subprime lending rates and foreclosure concentrations. This might explain why these areas recovered more quickly during the post-Recession period. Besides these race and class transitions, the authors indicate that the metropolitan context is important. Citywide data suggest that the Great Recession hit Chicago relatively hard compared to New York and Washington, DC, and this might help to explain the continued downward trajectory of the Bronzeville area. The argument resembles the findings of Newman and Goetz: when explaining neighborhood trajectories, attention should be paid not only to local developments, but also to developments at other spatial levels.

The paper by Katrin Großmann and Annegret Haase (2015, this issue) assesses the value of assemblage and complexity thinking for understanding urban neighborhoods. Their basic argument is that our thinking about neighborhoods is too linear in terms of decline or gentrification, and that we pay insufficient attention to differential developments within neighborhoods and through time (Hyra and Rugh also emphasize that this is important).

They argue that assemblage and complexity thinking allows us to better focus on such differential developments. As they state in their paper, assemblage thinking in urban research develops a view that claims to overcome reductionist, linear, causal thinking in favor of apprehending unexpected effects, shifts, and turns. They test whether such a different ontological perspective can enrich neighborhood change research by examining the development of an inner-city district and a large housing estate in Leipzig in the eastern part of Germany. Because it is necessary for their argument to give some elaborate descriptions of Leipzig and the two selected areas, the paper also provides some useful background information about a former Eastern European city and about the development of two areas that might be considered typical for such cities.

Does complexity and assemblage thinking help to better explain neighborhood developments? The authors conclude that it does, but not as a new "one-size-fits-all" approach that replaces all others. Assemblage thinking does, however, keep our eyes open to nuances of neighborhood developments that, at first sight, do not fit the story well. As the authors state: *"we might uncover unexpected dynamics, surprising differences, or counter-trends in that which seems to be stagnant or stable."* Such open thinking might indeed lead to rich results and new perspectives. In the case of developments in times of crisis, an assemblage approach may direct attention to developments that are not immediately related to economic developments, but that can influence the trajectory of (a part of) a neighborhood.

Roger Andersson and Lina Hedman (2016, this issue) begin their paper on Malmö (Sweden) with the hypothesis that the economic crisis is likely to be associated with increasing levels of income segregation and income polarization, and that poor neighborhoods are more severely hit by negative economic developments than more affluent ones. They investigated neighborhood developments in two different time periods – one which was characterized by a severe economic crisis and one in which the economy was relatively stable. Their findings indeed suggest that income segregation and income polarization increased during the economic crisis and that in this period poor neighborhoods fared worse than the region in general. The economic crisis of the early 1990s led to an overall increase in both income inequality and income segregation, but the already poor neighborhoods experienced more dramatic increases in unemployment rates and in relative share of low-income people than better-off neighborhoods. During the more economically stable period, these patterns and measures were also more stable. The authors state clearly that they cannot say that the economic recession *causes* these outcomes, but there is at least a clear correlation between these two. Negative developments can be a consequence of in situ changes (people in poor neighborhoods are more likely to lose jobs than those in other parts of the region) and of residential sorting, where the differences in income and employment status between in-movers, out-movers, and stayers are greater during the period of recession compared to the more stable period. This serves to underline the fact that, to understand neighborhood change, one should simultaneously examine the dynamics of in situ populations and those associated with residential mobility.

The final paper in this special issue, by Rebecca Tunstall (2016, this issue), shows us that neighborhood developments in the United Kingdom are generally slow: neighborhoods are more "slothful" than dynamic. Neighborhoods do change, e.g., under the influence of urban regeneration policies, or as a consequence of in situ changes or residential mobility (see the paper by Hedman and Andersson), but in general this does not radically alter their relative ranking in a city. For example, Tunstall (2016, this issue) cites evidence that the relative social status of neighborhoods in inner London in 1896 correlates highly with measures of deprivation for the same neighborhoods nearly a century later in 1991, and here own results, spanning 26 years, reveal a similarly slothful change. This also means that whilst policy measures, e.g., urban regeneration policies, may have effects (e.g., with respect to social cohesion or neighborhood reputation), these effects are often less marked than policy-makers might expect.

A final note

For many decades, in various countries, local and national governments, often in combination with private partners, have implemented policies with the aim of counter- ing neighborhood decline. This has been done under various headings, such as urban renewal, urban restructuring, social renewal, state-led gentrification, big cities policies, and many more. However, the economic crisis and subsequent austerity measures have led to a major policy shift and changing priorities (Zwiers et al., 2016, this issue). Neighborhoods and neighborhood decline are not high on the political agenda any- more. Implicitly, or sometimes even explicitly, governments point to the need for residents themselves to solve problems in their neighborhoods: they should be the main actors of neighborhood regeneration. However, when the problem consists of a decline in social cohesion and in the willingness to interfere with neighborhood developments, it is risky (or overly optimistic) to count on local residents or entrepre- neurs. Articles in this special issue indicate that the economic crisis, in combination with many other developments, does affect poor neighborhoods (and consequently the people living in these neighborhoods) more severely than more affluent ones. It is not efficient to stop supporting vulnerable neighborhoods, as it may nullify the (often long- term) investments that have been undertaken in the past in such neighborhoods. Moreover, it is neither realistic nor reasonable to expect residents to solve the problems in their neighborhoods. People in poor neighborhoods not only marshal their energies in their daily struggle to get by, they are often also faced with an environment with a high degree of cultural diversity, a lack of mutual trust between residents, a high level of turnover, and a high risk of being a crime victim. These are all ingredients that reduce the likelihood of people being, are willing, or able to intervene in their neighborhood in a positive way (see, e.g., Kleinhans & Bolt, 2014). Therefore, it is, in our view, the responsibility of (national and local) governments to prevent neighborhoods from sliding down to a level where the safety and health of their residents are compromised.

Acknowledgments

The guest editors of the special feature would like to thank all paper reviewers: Manuel Aalbers, Katrin Anacker, Hans Skifter Andersen, Philip Ashton, Annette Hastings, Jean Hillier, David Hulchanski, David Imbroscio, Dan Immercluck, Reinout Kleinhans, Mickey Lauria, Ruth Lupton, Thomas Maloutas, David Manley, Alan Murie, Fenne Pinkster, Olof Sjternstrom, Kiril Stanilov, Todd Swanstrom, Tiit Tammaru, and Sasha Tsenkova. We would also like to thank Deborah Martin for her excellent advice and great cooperation when preparing this special feature.

Disclosure statement

No potential conflict of interest was reported by the authors.

Funding

This work was supported by Platform31 (The Netherlands). Some of Maarten van Ham's time on this project was funded by the European Research Council under the European Union's Seventh Framework Programme (FP/2007–2013)/ERC Grant Agreement No. 615159 (ERC Consolidator Grant DEPRIVEDHOODS, Socio-spatial inequality, deprived neighbourhoods, and neighbourhood effects) and the Marie Curie Programme under the European Union's Seventh Framework Programme (FP/2007–2013)/Career Integration Grant No. PCIG10-GA-2011-303728 (CIG Grant NBHCHOICE, Neighbourhood choice, neighbourhood sorting, and neighbourhood effects).

References

Andersson, Roger, & Hedman, Lina (2016). Economic decline and residential segregation: A Swedish study with focus on Malmö. *Urban Geography*. doi: 10.1080/02723638.2015.1133993

Bolt, Gideon, Van Kempen, Ronald, & Van Ham, Maarten (2008). Minority ethnic groups in the Dutch housing market: Spatial segregation, relocation dynamics and housing policy. *Urban Studies*, *45*(7), 1359–1384.

Castells, Manuel (2000). *The information age: Economy, society and culture. Volume I: The rise of the network society* (2nd ed.). Malden, MA: Blackwell.

Cheshire, Paul (2012). Are mixed community policies evidence based? A review of the research on neighbourhood effects. In Maarten Van Ham, David Manley, Nick Bailey, Ludi Simpson, & Duncan Maclennan (Eds.), *Neighbourhood effects research: New perspectives* (pp. 267–294). Dordrecht: Springer.

Clark, William A.V., & Ledwith, Valerie (2006). Mobility, housing stress, and neighborhood contexts: Evidence from Los Angeles. *Environment and Planning A*, *38*(6), 1077–1093.

Ellen, Ingrid G., & Turner, Margery A. (1997). Does neighborhood matter? Assessing recent evidence. *Housing Policy Debate*, *8*(4), 833–866.

Enchautegui, Maria E. (1997). Latino neighborhoods and Latino neighborhood poverty. *Journal of Urban Affairs*, *19*(4), 445–467.

Fong, Erik, & Gulia, Milena (1999). Differences in neighborhood qualities among racial and ethnic groups in Canada. *Sociological Inquiry*, *69*(4), 575–598.

Friedrichs, Jürgen, Galster, George, & Musterd, Sako (2003). Neighbourhood effects on social opportunities: The European and American research and policy context. *Housing Studies*, *18* (6), 797–806.

Gans, Herbert J. (1962). *The urban villagers: Group and class in the life of Italian-Americans*. New York, NY: The Free Press.

Graham, Stephen, & Marvin, Simon (1996). *Telecommunications and the city: Electronic spaces, urban places*. London: Routledge.

Grigsby, William, Baratz, Morton, Galster, George, & Maclennan, Duncan (1987). The dynamic of neighborhood change and decline. *Progress in Planning*, *28*(1), 1–76.

Großmann, Katrin, & Haase, Annegret (2015). Neighbourhood change beyond clear story lines: What can assemblage and complexity thinking contribute to a better understanding of neighbourhood development? *Urban Geography*. doi: 10.1080/02723638.2015.1113807

Guest, Avery M., & Wierzbicki, Susan K. (1999). Social ties at the neighborhood level: Two decades of GSS evidence. *Urban Affairs Review, 35*(1), 92–111.

Hoyt, Homer (1939). *The structure and growth of residential neighborhoods in American cities.* Washington, DC: Federal Housing Administration.

Hyra, Derek, & Rugh, Jacob (2016). The US great recession: Exploring its association with black neighborhood rise, decline and recovery. *Urban Geography.* doi: 10.1080/02723638.2015.1103994

Kazepov, Yuri (Ed.). (2005). *Cities of Europe: Changing contexts, local arrangements, and the challenge to social cohesion.* Oxford: Blackwell.

Kleinhans, Reinout J., & Bolt, Gideon (2014). More than just fear: On the intricate interplay between perceived neighborhood disorder, collective efficacy, and action. *Journal of Urban Affairs, 36*(3), 420–446.

Lieberson, Stanley (1981). An asymmetrical approach to segregation. In Ceri Peach, Vaughan Robinson, & Susan Smith (Eds.), *Ethnic segregation in cities* (pp. 61–82). London: Croom Helm.

Logan, John, Stults, Brian, & Farley, Reynolds (2004). Segregation of minorities in the metropolis: Two decades of change. *Demography, 41*, 1–22.

Marcuse, Peter, & Van Kempen, Ronald (Eds.). (2000). *Globalizing cities: A new spatial order?* Oxford: Blackwell.

Musterd, Sako, Andersson, Roger, Galster, George, & Kauppinen, Timo (2008). Are immigrants' earnings influenced by the characteristics of their neighbours? *Environment and Planning A, 40*, 785–805.

Musterd, Sako, & Ostendorf, Wim (Eds.). (1998). *Urban segregation and the welfare state.* London: Routledge.

Musterd, Sako, & Van Kempen, Ronald (2009). Segregation and housing of minority ethnic groups in western European cities. *Tijdschrift voor Economische en Sociale Geografie, 100*(4), 559–566.

Newman, Kathe, & Goetz, Edward (2015). Reclaiming neighborhood from the inside out: Regionalism, globalization and critical community development. *Urban Geography.* doi: 10.1080/02723638.2015.1096116

Park, Robert E., Burgess, Ernest W., & McKenzie, Roderick D. (Eds.). (1925/1974). *The city.* Chicago, IL: Chicago University Press.

Prak, Niels L., & Priemus, Hugo (1986). A model for the analysis of the decline of postwar housing. *International Journal of Urban and Regional Research, 10*(1), 1–7.

South, Scott J., Crowder, Kyle, & Chavez, Erick (2005). Migration and spatial assimilation among U.S. Latinos: Classical versus segmented trajectories. *Demography, 42*, 497–521.

Stein, Maurice, R. (1972). *The eclipse of community.* Princeton, NJ: Princeton University Press.

Suttles, Gerald D. (1974). *The social order of the slum: Ethnicity and territory in the inner city.* Chicago, IL: The University of Chicago Press.

Taeuber, Karl E., & Taeuber, Alma F. (1965). *Negroes in cities.* Chicago, IL: Aldine.

Temkin, Kenneth, & Rohe, William (1996). Neighborhood change and urban policy. *Journal of Planning Education and Research, 15*(3), 159–170.

Tunstall, Rebecca (2016). Are neighbourhoods dynamic or are they slothful? The limited prevalence and extent of change in neighbourhood socio-economic status, and its implications for regeneration policy. *Urban Geography.* doi: 10.1080/02723638.2015.1096119

Van Ham, Maarten, & Manley, David (2010). The effect of neighbourhood housing tenure mix on labour market outcomes: A longitudinal investigation of neighbourhood effects. *Journal of Economic Geography, 10*, 257–282.

Van Kempen, Ronald, & Wissink, Bart (2014). Between spaces and flows: Towards a new agenda for neighbourhood research in an age of mobility. *Geografiska Annaler, Series B, Human Geography, 96*(2), 95–108.

Webber, Melvin M. (1964). The urban place and the non-place urban realm. In Melvin M. Webber, John W. Dijckman, Donald L. Foley, Albert Z. Guttenberg, William L.C. Wheaton, & Catherine Bauer Wurster (Eds.), *Explorations into urban structure* (pp. 79–153). Philadelphia, PA: University of Pennsylvania Press.

Wellman, Berry (1999). *Networks in the global village: Life in contemporary communities.* Boulder, CO: Westview Press.

Wellman, Berry (2001). Physical place and cyberplace: The rise of personalized networking. *International Journal of Urban and Regional Research, 25*(2), 227–252.

Zwiers, Merle, Bolt, Gideon, Van Ham, Maarten, & Van Kempen, Ronald (2016). Neighborhood decline and the economic crisis. *Urban Geography.* doi: 10.1080/02723638.2015.1101251

The global financial crisis and neighborhood decline

Merle Zwiers[a], Gideon Bolt[b], Maarten van Ham[a,c] and Ronald van Kempen[b]

[a]OTB – Research for the Built Environment, Faculty of Architecture and the Built Environment, Delft University of Technology, Delft, The Netherlands; [b]Faculty of Geosciences, Utrecht University, Utrecht, The Netherlands; cCentre for Housing Research, School of Geography and Geosciences, University of St Andrews, St Andrews, Scotland

ABSTRACT
Neighborhood decline is a complex and multidimensional process. National and regional variations in economic and political structures (including varieties in national welfare state arrangements), combined with differences in neighborhood history, development, and population composition, make it impossible to identify an ideal-type process of neighborhood decline over time. The recent global financial crisis and the subsequent economic recession affected many European and North American cities in terms of growing unemployment levels and rising poverty in concentrated areas. Investments in urban restructuring and neighborhood improvement programs have simultaneously decreased or come to a halt altogether. While many studies have investigated the effects of the financial crisis on national housing markets or on foreclosures in particular US metropolitan areas, only a few studies have focused on how the crisis affected neighborhood change. By proposing 10 hypotheses about the ways in which the economic crisis might influence processes of neighborhood decline, this article aims to advance the debate and calls for more contextualized, empirical research on neighborhood change.

Introduction

The global financial crisis, which started in 2008, has had a major impact on many Western European and North American countries. In the years preceding the crisis, many countries in the Global North experienced rising house prices, accompanied by an expansion of mortgage financing (Wachter, 2015). As financial networks have become increasingly global, the collapse of the subprime mortgage market and house price bubble in the United States has had repercussions on a global scale (Martin, 2011). While there were significant differences between impacted countries in the timing and macroeconomic processes underlying the financial crisis, the characteristics of the subsequent economic recession have been similar: stagnating economic growth, a sovereign debt crisis, and rising unemployment (Aalbers, 2015). Many governments have responded to the declining economy and growing unemployment levels with implementation of major budget cuts for social provisions (Peck, 2012). This has contributed to both relative and absolute growth in the number of economically disadvantaged households and has exacerbated poverty on both sides of the Atlantic. While the average income of the top 10% of the populations of OECD countries was essentially unaffected by the crisis, the average income of households in the lowest income decile experienced an annual decline of 2% between 2007 and 2010 (Organisation for Economic Co-operation and Development [OECD], 2013). In many countries, the financial crisis has also had a major impact on the housing market, evidenced by a large drop in home prices and declining sales of both existing and new-build housing (Van Der Heijden, Dol, & Oxley, 2011).

Today, many countries are slowly recovering from the first shocks of the financial crisis and the economic recession that followed. However, in many Southern European countries, unemployment rates continue to be very high and, although unemployment is declining in places like the United States and Germany, long-term unemployment appears to be a persistent problem in many countries (OECD, 2014; Shierholz, 2014). Similarly, despite gradual stock market recoveries and some modest increases in house prices, repercussions from the financial crisis and economic recession persist in all countries. In many countries, the crisis has had predictable effects on the supply side of the housing market – the willingness of banks to lend money to prospective owners has generally declined. In some countries, investors in real estate became more selective, avoiding projects with too much risk; in the United States, in contrast, investors of another ilk have bought large numbers of foreclosed, real estate owned (REO) proper-ties with the main goal of making a profit (e.g., Mallach, 2010b). Regeneration and restructuring initiatives have been put on hold throughout Western Europe (Boelhouwer & Priemus, 2014; Raco & Tasan-Kok, 2009; Schwartz, 2011). While some governments, such as the United Kingdom and the Netherlands, implemented stimulus programs to generate more (affordable) housing in the years after the crisis, recent budget cuts have put an end to these programs (Scanlon & Elsinga, 2014; Schwartz, 2011).

The demand side of the housing market has also changed. Banks have tightened lending terms, making it more difficult for many households to obtain a mortgage (Goodman, Zhu, & George, 2015). As a result, there is more demand for private rentals and social or public housing. The financial crisis has affected employment on both sides of the Atlantic, in terms of either high unemployment levels or a shift toward more casualized labor contracts such as zero hour or temporary employment contracts (Aalbers, 2015; Puno & Thomas, 2010). This has led to financial strain and housing affordability problems for many households (Joint Center for Housing Studies of Harvard University [JCHS], 2015). In the United States, households that are behind on their mortgage payments, and that are unable to obtain a mortgage modification with their lender, are faced with displacement due to fore-closure. This results in a large group of residents with badly damaged credit ratings who are in search of post-foreclosure housing in nearby areas (Martin, 2012). In other countries where the option of foreclosure is often not available, households that are unable to pay their rent or mortgage often have to move to cheaper dwellings and less attractive neighborhoods, while others have to stay put, because moving is too expensive or alternatives are not available, or because negative equity makes it impossible for them to move.

All of these developments have contributed to rising inequality in the Global North, particularly in terms of income and housing (for analyses of income inequality, see Bellmann & Gerner, 2011; Immervoll, Peichl, & Tatsiramos, 2011). The recent crisis therefore raises questions about the future development of neighborhoods, especially because inequality tends to have specific spatial outcomes including increased segregation, increased spatial concentration of low-income groups, and negative neighborhood effects (European Commission, 2010; Glaeser, Resseger, & Tobio, 2009; Van Eijk, 2010; Zwiers & Koster, 2015). While there has been little research on the effects of the crisis at the neighborhood level, the evidence described above suggests that the effects are distributed unevenly across urban areas (Batson & Monnat, 2015; Foster & Kleit, 2015). As households in the bottom income decile have experienced the sharpest drop in income, the effects of the crisis are likely to be felt most acutely in the most disadvantaged neighborhoods (see also Rugh & Massey, 2010; Thomas, 2013).

In view of these concerns, this article sets out to identify factors that affect neighborhood decline in the aftermath of the crisis. Many economists have pointed to structural changes in national housing markets and to the changing role of states as important consequences of the crisis; yet, few researchers analyze how these changes play out at the neighborhood level. Similarly, housing researchers have identified multiple drivers behind neighborhood decline, but many of them focus on within-neighborhood processes at the expense of developments at higher scales (Van Beckhoven, Bolt, & Van Kempen, 2009). In this article, we aim to bridge this gap by presenting 10 hypotheses on how changes at different geographical scales affect neighborhood decline. Our goal is not to create the next ideal-type model of neighborhood decline processes; rather, we seek to further the intellectual debate on neighborhood decline and call for more research on the spatial consequences of the crisis, specifically on neighborhoods as an important territorial dimension of increasing inequality.

Our hypotheses mainly pertain to the Global North. Although these countries have very different political, economic, and social structures, research on neighborhood change in different contexts in the Global North has often led to broadly similar findings. This suggests that many of the underlying processes of neighborhood change are comparable across countries. In the same vein, the increasingly global nature of financial and housing markets (Aalbers, 2015) yields similarities in the effects of the financial crisis and the economic recession between countries. However, the effects of the global financial crisis are mediated by national policies, local (housing market) circumstances, and intra-neighborhood processes, meaning that the crisis has different outcomes in different places.

The next section of this article presents a short discussion of definitions of neighborhoods and neighborhood decline. We then highlight important elements from existing studies to formulate 10 hypotheses about the effects of the financial crisis and of the economic recession on neighborhood decline. These hypotheses are divided over three sections, each with a different geographical focus. The conclusion brings our arguments together and calls for more contextualized longitudinal research.

Defining neighborhoods and neighborhood decline

Neighborhoods are defined in various ways. Some definitions are related to distance: the neighborhood covers the area within which one can reach important destinations (schools, shops, and friends) within walking distance (e.g., Morris & Hess, 1975). Other definitions are based on social networks and refer explicitly to the existence of social bonds in the area (e.g., Warren, 1981). However, these definitions imply that "the neighborhood" is different for each individual, which makes research on neighborhood outcomes extremely complicated. George Galster defines neighborhoods as "… bundles of spatially based attributes associated with clusters of residences, sometimes in conjunction with other land uses" (Galster, 2001, p. 2112). The "spatially based attributes" refer to, for example, the characteristics of buildings, as well as infrastructural, demographic, class, status, social interactive, and sentimental characteristics. Defining neighborhoods based on spatial similarities (such as housing type or population composition) is difficult, especially in mixed-housing areas.

All definitions of neighborhoods have their advantages and disadvantages and there is no ideal neighborhood definition. The choice of definition depends on the type of research and should be substantiated by the researcher, bearing in mind that different definitions of neighborhoods may lead to different outcomes. For our purposes, it is sufficient to use a rather general and pragmatic definition of neighborhood: a neighborhood is a relatively small spatial subdivision of a city or town for which a number of physical, demographic, and socioeconomic characteristics can be measured. The size of a neighborhood may vary by city.

Neighborhoods play an important role in shaping the lives of individuals and households, in relation to social contacts, identity, health, and happiness (see also Martin, 2003). Moreover, neighborhoods have become increasingly important as local political and economic entities, with many governments focusing on neighborhoods to solve a wide array of social and economic problems (Martin, 2003). This highlights the importance of neighborhoods in a post-crisis society: with declining national government involvement in many countries, there may be an even stronger need to deal with many problems locally, on, for example, the level of cities or neighborhoods.

Neighborhoods can develop in different directions: a neighborhood can be demographically stable for years or even decades. Neighborhoods can experience gentrification, indicated by, for example, rising house prices, an outflow of low-income households, and an inflow of more affluent households. The extensive literature on this topic documents such processes in great detail (e.g., Doucet, 2014; Lees, 2008). Neighborhoods can also show a process of decline, indicated by falling house prices, an inflow of low-income households and an outflow of more affluent households.

In this article, we assume that the long-lasting effects of the global financial crisis and the economic recession will fuel neighborhood decline. We use a broad definition of neighborhood decline: any negative development in the physical, demographic, or socioeconomic conditions of a neighborhood as experienced by its residents or other stakeholders.

Ten hypotheses on the crisis and neighborhoods

The remainder of this article consists of 10 hypotheses about the ways in which the crisis might influence neighborhood decline. They are intended as a challenge to researchers to test whether these hypotheses can be confirmed or rejected in different national and urban contexts. The hypotheses are divided into three sections. The first part focuses on how the crisis plays out in national housing and welfare systems. The next part zooms in on the local context as a mediating variable in processes of neighborhood decline, while the final part concentrates on residents as drivers of neighborhood change.

Part 1: The role of national housing and welfare systems

Differences in welfare state regimes are an important explanatory factor in the wide range of national differences in housing systems (Priemus & Whitehead, 2014). In countries where the government has historically been strongly involved in the devel-opment of affordable (social) housing, such as Denmark, Sweden, and the Netherlands, the quality and the size of the social housing stock was originally very high (Tsenkova & Turner, 2004; Van Kempen & Priemus, 2002). This high initial quality has mitigated processes of neighborhood decline and has led to relatively low levels of income segregation in these countries. However, over the past few decades, severe cuts in housing subsidies took place in these countries and they have moved toward a more market-based housing system, where the responsibility for social housing shifted from public authorities to housing associations or NGO landlords. Housing associations are now increasingly dependent on their own revenue to construct new social housing (Schwartz, 2011; Van Kempen & Priemus, 2002). To generate revenue, many housing associations have been selling off the better parts of their social housing stocks over the past decade, significantly reducing the size and average quality of the social housing stock (Kleinhans & Van Ham, 2013; Schwartz, 2011).

In many countries, the financial crisis has led to the implementation of budget cuts and austerity programs. In combination with cuts in (social) housing subsidies before the crisis, these austerity programs have had an important impact on the opportunities for households on the housing market. Firstly, especially in times of economic reces-sion, austerity programs and budget cuts directly affect the financial resources of households (cf. Lindbeck, 2006; Swank, 1998). Secondly, austerity programs and budget cuts have restricted the resources available for the maintenance and construction of affordable social housing, although these processes have been more dramatic in some countries than in others (Priemus & Whitehead, 2014; Van Der Heijden et al., 2011). In the United States, for example, Low-Income Housing Tax Credit (LIHTC) programs were implemented in the 1980s and these programs were extended during the mortgage crisis and the years after to stimulate the development of low-income housing (Schwartz, 2011). However, because of the unstable market for tax credits, the LIHTC program tends to be more successful in the more robust housing markets in major metropolitan areas where banks are still dependent on the Community Reinvestment Act (Belsky & Nipson, 2010; Schwartz, 2011). Next to showing geographical differences in the effectiveness of tax credit programs, it is unlikely that they will generate as much equity for housing as it did before the crisis (Schwartz, 2011).

We can thus see that the financial crisis has affected the production of affordable housing in many countries in different ways. In countries where housing associations are dependent on private investors, we can expect to see the production of social housing to increase in those areas where there is a more robust housing market and where there is potential for financial gain. In other countries, we can generally expect a declining production of affordable housing. Together with more financial restrictions for households as a direct effect of the crisis, these processes can reduce residential mobility and force low-income groups to concentrate in neighborhoods where affordable housing options are still available. This can easily lead to increasing concentrations of low-income groups in the most deprived areas.

> Hypothesis 1: Austerity programs and budget cuts lead to a smaller social safety net for vulnerable groups on the one hand, and to more limited options on the social housing market on the other, leading to increasing concentrations of low-income groups in particular neighborhoods.

The extent of the impact of the crisis on the housing market depends on the volatility and structure of local and regional housing markets in different countries (Van Der Heijden et al., 2011). In countries with highly regulated housing finance systems, such as Germany, Switzerland, and Austria, the housing market was barely affected by the crisis (Whitehead, Scanlon, & Lunde, 2014). The most important explanations for housing market stability in these countries are the well-developed rental markets and the low homeownership rates, together with conservative lending policies (Schneider & Wagner, 2015; Whitehead et al., 2014). In countries with more open finance markets, of which Ireland and Iceland are the main examples, house prices fell considerably due to the rapid expansion of mortgage debt in the years before the crisis (Whitehead et al., 2014).

In countries with high mortgage indebtedness, states and financial institutions deliberately stimulated homeownership over the past few decades. First, many low- to middle-income groups and first-time buyers were allowed to obtain a mortgage by engaging in high loan-to-value lending (Schelkle, 2012). Second, direct subsidies or tax allowances were implemented to support low- to middle-income groups' entry into homeownership (though in some countries, subsidies such as mortgage interest deductions tend to benefit high-income groups the most) (Hanson, Brannon, & Hawley, 2014; Schelkle, 2012). Low- to middle-income groups have generally been hit the hardest by the financial crisis and the subsequent economic recession in terms of underwater mortgages, unemployment, and declining incomes (Dreier, Bhatti, Call, Schwartz, & Squires, 2014).

In the United States, subprime and predatory lending practices have disproportionally targeted disadvantaged groups in disadvantaged neighborhoods (Aalbers, 2009; Martin, 2011; Mayer & Pence, 2008). Subprime and predatory lending generally refer to high loan-to-value lending, compensating for higher credit risks with unfavorable terms such as higher fees and interest rates that are not beneficial to the borrower (Aalbers, 2013; Crossney, 2010). These practices increase the debt of the borrower beyond the collateral property and reduce the value of the underlying asset and accumulated equity (Crossney, 2010; Schloemer, Wei, Ernst, & Keest, 2006). Subprime and predatory lending tended to be spatially clustered in particular disadvantaged and segregated parts of the US cities, resulting in high numbers of foreclosures in these areas (Anacker & Carr, 2011; Batson & Monnat, 2015; Crossney, 2010; Hyra & Rugh, 2015; Immergluck, 2008; Mallach, 2010a; Rugh & Massey, 2010). Concentrations of foreclosures and vacancies in particular areas may lead to declining house values of nearby properties (Immergluck & Smith, 2006; Immergluck, 2009) and fuel neighborhood decline through vandalism and increasing crime rates (Aalbers, 2013; Jones & Pridemore, 2014; Martin, 2011; Newman, 2009; Ojeda, 2009).

In general, declining house prices have disproportionally affected low- to middle-income groups, often leaving them with a very unstable financial situation and negative equity (Crossney, 2010; Dreier et al., 2014; Thomas, 2013). In the United States, this has resulted in high concentrations of foreclosures in disadvantaged neighborhoods, displacing large numbers of people who are in need of (affordable) housing and have lost the ability to obtain a mortgage due to badly damaged credit (Goodman et al., 2015; Martin, 2012). These post-foreclosure households tend to relocate in other hard-hit foreclosure areas, contributing to declining average household income and neighborhood instability (Martin, 2012).

Hypothesis 2: The neighborhood effects of the crisis are stronger in countries that have actively stimulated homeownership at high loan-to-value rates. Vulnerable groups such as racial or ethnic minorities, low- to middle-income households, and first-time buyers are especially affected by the crisis. When these groups are overrepresented, in particular neighborhoods, these neighborhoods are often affected by rapid processes of decline.

In countries where there has been a deliberate policy to expand homeownership over the past few decades, it has become more difficult for low- to middle-income groups and first-time buyers to obtain a mortgage than in the years preceding the crisis (Boelhouwer & Priemus, 2014; Clark, 2013; Goodman et al., 2015). The mortgage systems that have emerged from the crisis generally favor higher-income groups, leading to increasing disparities between financially stable and financially unstable households (Forrest & Hirayama, 2015). This ultimately means that particular groups and areas are excluded from the mortgage housing market (Clark, 2013; Forrest & Hirayama, 2015; Martin, 2011; Watson, 2009). When it is more difficult for low- to middle-income groups to obtain a mortgage, they are forced to turn to the rental sector. Because renters tend to spend a significantly higher share of their income on housing costs than homeowners (e.g., Haffner & Boumeester, 2014) and because they are not able to accumulate housing equity, this will ultimately contribute to increasing inequality between renters and owners.

Hypothesis 3: After the crisis, low- to middle-income groups and first-time buyers are increasingly excluded from the mortgage market, which creates a large group in need of affordable rental housing. At the same time, these changes will lead to a declining home-ownership rate in particular areas, creating a spatial divide based on different tenures, and ultimately leading to increasing inequality.

Housing opportunities typically differ between generations. The recent crisis and subsequent recession is likely to further increase intra-generational differences in terms of housing opportunities (e.g., Forrest & Hirayama, 2015). There is already a clear difference between older generations and younger generations – the former have been more able to transform their housing investments into assets over time. High student debts, long-term unemployment, a shift toward a more casualized workforce and stricter mortgage eligibility criteria make it more difficult for the millennial generation (born 1985–2004) to pursue homeownership (JCHS, 2015). The older members of this cohort are just entering the housing market and studies have shown that only a small percentage has been able to become homeowners; this is even more difficult for minority groups (Clark, 2013; JCHS, 2015). In many countries, there has been a decline in homeownership rates among younger households as they postpone marriage and childbirth and tend to prolong their stay in the parental home (Aalbers, 2015; JCHS, 2015; Lennartz, Arundel, & Ronald, 2015).

Although many young people might have always been dependent on family financial support to some extent (in the sense of receiving down payments), in recent times, the dependence on family resources to achieve homeownership is becoming more pronounced (Forrest & Hirayama, 2015). However, as many parents have also been subjected to the effects of the financial crisis and the recession (in terms of unemployment, declining incomes, foreclosures, and negative equity), parents are not equally able to transfer wealth to their children. This is especially true for younger, lower educated, and minority groups that have accumulated only modest equity (Clark, 2013). In the long run, children from more privileged families will be able to maintain their relatively privileged status by investing in homeownership and accumulating wealth through mortgage amortization and housing appreciation (Forrest & Hirayama, 2015; Rohe, Van Zandt, & McCarthy, 2002). Children from more economically deprived backgrounds, however, will be more dependent on the rental market, thereby increasing their housing costs and reducing their ability to use homeownership as a way to accumulate wealth. These developments will ultimately lead to strong inter- and intra-generational disparities on the housing market (see also Clark, 2013; Forrest & Hirayama, 2015).

> Hypothesis 4: The crisis has fueled intra-generational differences in terms of housing opportunities. This will increase the influence of social class and inter-generational transmission of resources as stratifying factors.

Countries like Japan, England, the United States, and Australia witness an increase in the proportion of households (often young people) who enter the private rental sector (Forrest & Hirayama, 2015). There is much concern amongst scholars that the rise of the private rental sector has negative consequences for both the renters and the neighborhoods in which these houses are concentrated. In the United States, for example, the number of foreclosed properties owned by banks and other mortgage lenders has spiked in the post-crisis period. These REO properties are often acquired by private investors with the main goal of making their investment profitable (Mallach, 2010b). Scholars and activists fear that investors in private housing have little interest in maintaining these dwellings and that practices of "milking" and speculation will spur the process of neighborhood decline (Aalbers, 2013; Ellen, Madar, & Weselcouch, 2014; Fields & Uffer, 2014; Forrest & Hirayama, 2015).

Although the US federal government has invested billions into the Neighborhood Stabilization Program targeting REO and other vacant properties, the majority of these properties are purchased by private investors rather than owner-occupiers (Ellen et al., 2014). Researchers have argued that private investors play an important role in reducing concentrations of REO properties in particular neighborhoods and that they have been successful in reducing vacancy periods (Ellen et al., 2014; Immergluck, 2010; Pfeiffer & Molina, 2013). Despite the widespread assumption that the sale of REO properties to private investors accelerates neighborhood decline in the most hard-hit neighborhoods due to a lack of maintenance (e.g., Mallach, 2010a), recent studies show that not all private investors adopt business models that negatively affect neighborhoods (Ellen et al., 2014; Immergluck & Law, 2014; Mallach, 2010b).

Though corporate investment does not necessarily harm neighborhoods, the conversion of REO properties into rental units might still fuel processes of neighborhood decline. First of all, renting out properties can contribute to neighborhood instability because of high turnover rates (Kleinhans & Van Ham, 2013; Mallach, 2010a). Second, research has shown that properties sold to private investors and converted into rental units negatively affect the value of surrounding properties (Ihlanfeldt & Mayock, 2014).

> Hypothesis 5: The crisis has led to an increase of corporate investment in the private rental sector. Converting properties into rental units might lead to neighborhood instability and might negatively impact surrounding property values. These effects will be the strongest in the most hard-hit neighborhoods and are likely to have negative spillover effects on surrounding areas.

Part 2: The mediating role of the local context

The effects of the crisis and recession, and the austerity programs and budget cuts that followed, are unevenly distributed within countries (cf. Peck, 2012). Cities have been hit hardest, because housing markets are essentially localized and public services and social housing generally tend to be concentrated in city areas (Blank, 1988; Borjas, 1999; Peck, 2012). Yet, the effects of the crisis differ between cities. Although most scholars have focused mainly on neighborhood-level characteristics to explain neighborhood decline, Jun (2013) argues that metropolitan and municipal factors significantly affect neighborhood change. Jun (2013) finds that neighborhood economic status trends in a positive direction in smaller and more homogeneous cities (in terms of race/ethnicity), while the reverse applies to larger heterogeneous cities. Her explanation is that smaller cities are less bureaucratic, that there is more room for citizen participation, and that the spending on public goods is lower in ethnically and racially diverse cities, possibly because there are more dissenting views than in homogeneous cities (Jun, 2013).

At the metropolitan level, economic strength is obviously an important factor associated with neighborhood change. Lauria and Baxter (1999) showed how the economic shock in New Orleans in the 1980s (caused by falling oil prices) led to the racial transition of neighborhoods, through the mechanisms of foreclosures. It intensified White flight from neighborhoods with relatively small but increasing Black populations. While Lauria and Baxter (1999) focused on the effect of a regional economic downturn, Hyra and Rugh (2015, this issue) look at the effects of the Great Recession that followed the global financial crisis. They compare three gentrifying African American neighborhoods in Chicago, New York, and Washington, DC. The Chicago neighborhood suffered more than the other two from foreclosure and house price decline, whereas the home values in the other two neighborhoods have recovered to pre-recession levels. This may be related to the fact that the recession hit Chicago relatively hard, which led to a higher unemployment and vacancy rate than in the other two cities.

> Hypothesis 6: The crisis has the strongest negative effect on neighborhoods in metropolitan areas with a weak economy and their recovery (if any) will also take longer than in neighborhoods that are situated in a strong regional economy.

In addition to exogenous factors that can set off processes of neighborhood decline, some of which we have identified above, characteristics of the neighborhood itself may fuel or mitigate these processes. The initial economic status of a neighborhood is a very strong predictor of its course of development in the long run. Meen and colleagues (2013) have shown how some areas have always had a natural advantage over others because of their location and/or access to particular resources, such as a proximity to ports or transportation centers, and that they maintain their high-quality status and position in the neighborhood hierarchy over longer periods of time.

The importance of the relative "starting position" of a neighborhood also relates to the physical quality of the dwellings. Some authors take an almost deterministic stance regarding the relevance of this "hard" variable (e.g., Newman, 1972; Coleman, 1985; and to a lesser extent Power, 1997). In the European context, there is much research on neighborhoods with a high share of post-war, high-rise residential buildings that are prone to processes of neighborhood decline due to the low quality of, and technical problems with, these buildings (Dekker & Van Kempen, 2004; Kearns, Whitley, Mason, & Bond, 2012; Kleinhans, 2004; Prak & Priemus, 1986; Van Beckhoven et al., 2009). But also in the US context, high foreclosure rates and predatory lending practices cannot only be attributed to the socioeconomic profile of residents (Strom & Reader, 2013). Neighborhoods characterized by a marginal housing stock and poor residents are often explicitly targeted by investors looking to make a profit (Aalbers, 2006; Strom & Reader, 2013)

However, the position of neighborhoods in the neighborhood hierarchy is not only a question of location or physical quality, but also a consequence of social processes. Similar types of housing (in physical terms) can acquire a vastly different social status dependent on the identity of a neighborhood. This identity can be very long-lasting (see Tunstall, 2015, this issue). Comparing three neighborhoods in Stirling, Scotland, Robertson and colleagues (2010) show that the social positioning in terms of class (poor, "respectable" working class, and middle class) did not significantly change since the time these neighborhoods were built (1920s and 1930s). This reveals that neighborhood reputations are sticky, which is partly due to the one-sided way in which neighborhoods are covered in the local media (Kearns, Kearns, & Lawson, 2013; see also Tunstall, 2015). Similarly, Loïc Wacquant (2008) has shown how political and academic debates on the American ghetto reinforce divisions in society based on race and class, thereby contributing to collective processes of stigmatization and exclusion. The stigmatizing perception of neighborhoods with concentrations of poor and/or racial/ethnic minorities as disordered environments leads to a reinforcement of segregation as middle-class residents, and especially Whites, are moving (or staying) away from these kinds of neighborhoods (Sampson, 2009).

Hypothesis 7: Areas that are characterized by a low-quality housing stock and a negative reputation are particularly prone to processes of neighborhood decline.

Over the past several decades, many countries have implemented neighborhood regeneration programs. The general goal of these programs was to reduce relative inequality between the most disadvantaged neighborhoods and the city or the national average (Jivraj, 2012). The ways in which these urban renewal programs are pursued in practice differ between countries (Skifter Andersen, 1999). However, in general, policies were implemented to stimulate a socioeconomic residential mix in deprived neighborhoods. Examples are the HOPE VI program in the United States, the Urban Restructuring Program in the Netherlands, and the New Deal for Communities in the United Kingdom (Bolt & Van Kempen, 2011; Goetz, 2010; Phillips & Harrison, 2010).

Many policymakers believe that the mixing of different socioeconomic groups in disadvantaged areas will lead to neighborhood upgrading (Andersson & Musterd, 2005; Van Gent, Musterd, & Ostendorf, 2009). In many cases, urban regeneration meant the demolition of low-quality rental dwellings, replacing them with more upmarket owner-occupied and luxury rental dwellings (Kleinhans, 2004). In this way, spatial concentrations of low-cost rental dwellings were reduced and the residents of the demolished dwellings were forced to relocate to other (often nearby, often also disadvantaged) neighborhoods where affordable housing was still available (Bolt, Van Kempen, & Van Ham, 2008; Crump, 2002; Posthumus, Bolt, & Van Kempen, 2013; Van Kempen & Priemus, 2002). Most of these residents did not have the means to move back to the more expensive, newly created housing in the regeneration area (Kleinhans & Varady, 2011). It has thus been argued that regeneration programs may lead to the downgrading of other (surrounding) neighborhoods, because the previous spatially concentrated deprivation becomes dispersed over a larger geographical area (Andersson, Brämå, & Holmqvist, 2010; Brama, 2013; Posthumus et al., 2013).

While these mixing policies can be successful in improving the economic statistics of a neighborhood, most of these policies have been heavily criticized for failing to really improve the lives of the original residents (Doff & Kleinhans, 2011; Goetz, 2010; Van Ham & Manley, 2012). Nevertheless, policymakers often frame such programs as successful, and these programs have contributed to some extent of segregation decrease (Feins & Shroder, 2005; Frey, 2010; Musterd & Ostendorf, 2005). A well-known argument is that countries such as the Netherlands and Sweden do not have ghetto-like neighborhoods *because* of a strong government involvement and mixing policies. This raises the question whether the retreat of governments from deprived neighborhoods as a result of the crisis will fuel processes of socioeconomic segregation and neighborhood decline. On the basis of Tunstall's (2015) conclusion that neighborhood renewal policies have not made a significant change in the neighborhood hierarchy, one might speculate that government retreat does not make much of a difference. On the other hand, it can be argued that whether a neighborhood is at the bottom rung of the ladder is not the only important factor; stratification between neighborhoods also contributes to their various fates.

Hypothesis 8: The crisis will have the largest effect on processes of neighborhood decline in neighborhoods where there has been strong government involvement in urban regeneration and other neighborhood policies.

Part 3: Behavioral responses: exit and voice

The dynamics of a neighborhood are highly affected by the decisions of its residents. Following Hirschman's (1970) "exit, voice and loyalty" framework, Permentier and colleagues (2007) argue that residents who are dissatisfied with their neighborhood can either choose to move out (exit) or adopt problem-solving strategies (voice). Loyalty (the attachment to neighborhood and its residents) increases the likelihood of the voice option and reduces the probability of residential mobility (Permentier et al., 2007).

Residential mobility is the central explanatory variable in the neighborhood decline model of Grigsby and others (1987). Neighborhoods can change rapidly as a result of selective mobility where the demographic and socioeconomic characteristics of those households leaving are different from the characteristics of the newly arriving households. Declining housing and neighborhood quality can spur residential mobility: middle- and higher-income groups move away from declining neighborhoods as a result of the decreasing attraction of dwellings and neighborhoods and the creation of new dwellings elsewhere – a process also known as relative depreciation (Grigsby et al., 1987; Hoyt, 1939). The likelihood of moving depends on whether household preference can be realized by the resources available to the household within the opportunities (available dwellings) and restrictions (ability to obtain a mortgage) of the desired housing market (Clark & Dieleman, 1996; Mulder & Hooimeijer, 1999). Generally speaking, more affluent households have a larger choice set of dwellings and neighborhoods.

The financial crisis and subsequent recession is likely to have major impacts on residential mobility. On the one hand, we have argued that people tend to be more limited in their options due to financial restrictions and stricter mortgage eligibility criteria. Households might *want* to move, but *are not able* to move because they cannot obtain a mortgage or do not find a suitable rental dwelling. In the European context, many low-income households are dependent on the availability of social or public housing and waiting lists are long, making it difficult for these households to move from one to another rented dwelling. Similarly, many homeowners in Western Europe might be forced to stay in a particular dwelling and neighborhood, because they cannot sell their current home without taking a large financial loss.

In the US context, foreclosures force people to move and thus lead to a wave of residential moves at first. However, the unstable financial situation of many foreclosed households, together with tight credit standards, makes it nearly impossible for these households to obtain a mortgage in the future (Goodman et al., 2015; Martin, 2012). Residential mobility can therefore also be expected to decrease in the United States, although a recent study by Pfeiffer and Molina (2013) shows how the foreclosure crisis offers an opportunity for Latino households in terms of socioeconomic mobility; however, they also argue that Latino households are more likely to purchase properties in Latino-concentrated areas, thereby exacerbating existing patterns of spatial segregation (Pfeiffer & Molina, 2013). Similarly, research has shown how many fore-closed households tend to end up in other hard-hit foreclosure areas (Martin, 2012), after which they are more or less stuck in these neighborhoods because they are unable to obtain a mortgage and move to a different area.

The unstable financial situation of many households, combined with stricter mort-gage eligibility, complicates residential mobility on both sides of the Atlantic. Even though residential mobility has decreased on both continents, the outcomes may be very different. In the United States, we can expect that limited residential mobility has further contributed to existing socioeconomic and racial segregation, while in Europe, it can be expected that the process of segregation has slowed down.

Hypothesis 9: Decreases in residential mobility rates can have different outcomes in different contexts. In many Western European countries, we expect a lower likelihood of an increase in residential segregation, while in the United States, foreclosures have led to a small short-term upsurge in residential mobility patterns, exacerbating existing segregation.

If residents are not satisfied with their neighborhood, they can (instead of moving out) also opt to organize themselves to address neighborhood problems. Whether that is a feasible strategy depends on the level of social cohesion in the neighborhood. It is often assumed that disadvantaged neighborhoods suffer from the lack of strong social ties and the advantages these ties bring along (Forrest & Kearns, 2001). Without a strong social fabric, neighborhoods are more prone to disorder in terms of vandalism, nuisance, and crime (Kleinhans & Bolt, 2014). Social disorganization theory, which originated from the Chicago School of Sociology, stated that disorganization in neigh-borhoods is caused by incapability of the local community in terms of a lack of (access) to resources, residential instability, or a weak social network (Shaw & McKay, 1942). Physical and social problems arise because residents are not able to enforce certain norms and to maintain social control. As a result, governments tend to retreat from public space and residents lose their trust in each other and "hunker down" (Putnam, 2007; Ross, Mirowsky, & Pribesh, 2001). Some researchers have argued that small levels of disorder (such as graffiti or broken windows) give rise to more serious crime offenses. The broken windows theory states that potential criminals interpret these levels of disorder as a sign of a lack of social control or involvement of the residents, and as such, feel free to engage in criminal behavior (Wilson & Kelling, 1982).

Recent research by Jones and Pridemore (2014) on the effect of vacancies on crime rates after the crisis concludes that population loss and vacant homes complicate neighborhood social organization. In line with social disorganization theory and the broken windows theory, they argue that the lack of collective efficacy as a result of low levels of population density makes those areas more attractive to criminals (Jones & Pridemore, 2014). In times of crisis, social cohesion in (disadvantaged) neighborhoods can develop in different directions. With many governments retreating, an increasing level of responsibility for the neighborhoods has shifted to its residents. In these neighborhoods, where many residents are unable to move, people may feel close to each other because of a common fate, actually increasing social cohesion. This can lead to a strengthening of solidarity networks and a deepening attachment to place, even in very stigmatized areas like the French banlieues (Kirkness, 2014). However, it is also possible that neighborhoods experiencing an inflow of lower-income groups are prone to increasing social disorganization. A change of population composition might lead to residential stress as people tend to prefer a neighborhood population that matches their own characteristics (Feijten & Van Ham, 2009; McPherson, Smith-Lovin, & Cook, 2001).

> Hypothesis 10: In times of crisis, social cohesion may be reinforced in areas where there has been a reasonable level of social interaction in the past, while it is likely to crumble in areas that experience increasing tensions because of a diversification of the population, or in areas that are experiencing significant declines in population density.

Conclusions

In this article, we have argued that contemporary neighborhood decline is a multi-dimensional process fueled by several macroeconomic processes related to the financial crisis and the recession that followed. However, we have also argued that there are several local and internal factors that might function as a mediating factor in processes of neighborhood decline. The interaction of micro-, meso- and macro-level factors heavily depends on the context in space and time.

There is a lack of empirical studies focused on the effects of the financial crisis on neighborhoods and their residents. In an attempt to push the debate forward, we have formulated 10 hypotheses on how the crisis might interplay with processes of neighbor-hood decline. We submit these hypotheses as a guide for future empirical research. Research is necessary because differences in the local effects of the crisis are likely to lead to a widening of the gap between wealthy and disadvantaged neighborhoods, between high-income mortgage borrowers and low-income borrowers, between privi-leged and less privileged households, and between renters and homeowners (Forrest & Hirayama, 2015). In combination with severe budget cuts and the implementation of austerity programs, this raises concerns about increasing spatial segregation based on social class (see also Tammaru, Marcińczak, Van Ham, & Musterd, 2015).

We have identified several factors from the literature that influence neighborhood change. However, little is actually known about the ways in which these factors interact in different contexts. We therefore call for more longitudinal research of neighbor-hoods *and* households that focuses on the drivers of neighborhood decline and disinvestment and, more generally, neighborhood change. Without longitudinal data on the residential and social mobility of households, it is difficult to disentangle the relative weight of residential sorting and incumbent processes in explaining neighborhood change. Incumbent upgrading and downgrading refers to the changing socioeconomic profile of the resident population within an area (e.g., Teernstra, 2014). It is an empirical question regarding how important external forces and internal developments are to neighborhoods; this can differ by country, city, or even by neighborhood.

This question is crucial, especially because in countries where the crisis has reduced residential mobility, incumbent processes may become relatively more important in explaining neighborhood decline through processes of rising unem-ployment and declining incomes (Andersson & Hedman, 2015). Individual-level data over long periods of time are needed to address this question. Such data are not available in all countries; however, as better data becomes available, researchers should aim to take a richer array of longitudinal individual and spatial variables into account (Van Ham & Manley, 2012). This is not only an academic question, but also relevant in the evaluation of neighborhood regeneration programs. Is there, for instance, an improvement in the livability and social status of neighborhoods due to the empowerment of the sitting population or due to the replacement of vulnerable groups by middle-class households?

Most studies that focus on neighborhood change tend to concentrate on case studies of specific cities, or specific gentrifying or declining neighborhoods. This focus can be largely attributed to the complexity of the subject, a lack of detailed (comparable) longitudinal data, and a bulk of statistical problems with which researchers are confronted; it nevertheless constitutes a large gap in research on neighborhood dynamics. Neighborhoods do not operate in a vacuum and while a particular neighborhood may experience absolute change, the picture may be completely different when we look at the relative change in a city or a country. Moreover, in a globalizing world, with growing, internationally connected economies and housing markets, it will become increasingly important to understand neighborhood change from a global perspective.

The financial crisis has had different economic, physical, social, and health-related outcomes, most of which we are only now beginning to grasp. Researchers have argued that the crisis has had different local outcomes between *and* within countries (Aalbers, 2009), but we have little insight into the long-term effects of the crisis on neighborhoods and its residents. It is important to understand how the crisis has affected spatial patterns of increasing inequality and neighborhood trajectories. A deeper understanding of the drivers behind neighborhood decline can contribute to the development of effective policymaking in the aftermath of the financial crisis and the economic recession.

Acknowledgment

We would like to thank Deborah Martin (*Urban Geography*) and the anonymous reviewers for their useful comments that greatly contributed to improving the article.

Disclosure statement

No potential conflict of interest was reported by the authors.

Funding

The research leading to these results has received funding from Platform31 in the Netherlands; from the European Research Council under the European Union's Seventh Framework Programme (FP/2007-2013)/ERC Grant Agreement number 615159 (ERC Consolidator Grant DEPRIVEDHOODS, Socio-spatial inequality, deprived neighborhoods, and neighborhood effects); and from the Marie Curie programme under the European Union's Seventh Framework Programme (FP/2007-2013)/Career Integration Grant number PCIG10-GA-2011-303728 (CIG Grant NBHCHOICE, Neighborhood choice, neighborhood sorting, and neighborhood effects).

References

Aalbers, Manuel B. (2006). 'When the banks withdraw, slum landlords take over': The structuration of neighbourhood decline through redlining, drug dealing, speculation and immigrant exploitation. *Urban Studies, 43*(7), 1061–1086.

Aalbers, Manuel B. (2009). Geographies of the financial crisis. *Area, 41*(1), 34–42.

Aalbers, Manuel B. (2013). How do mortgage lenders influence neighborhood dynamics? Redlining and predatory lending. In Maarten Van Ham, David Manley, Nick Bailey, Ludi Simpson, & Duncan MacLennan (Eds.), *Understanding neighborhood dynamics: New insights for neighborhood effects research* (pp. 63–85). Dordrecht: Springer Netherlands.

Aalbers, Manuel B. (2015). The great moderation, the great excess and the global housing crisis. *International Journal of Housing Policy, 15*(1), 43–60.

Anacker, Katrin B., & Carr, James H. (2011). Analysing determinants of foreclosure among high-income African–American and Hispanic borrowers in the Washington, DC metropolitan area. *International Journal of Housing Policy, 11*(2), 195–220.

Andersson, Roger, Brämå, Åsa, & Holmqvist, Emma. (2010). Counteracting segregation: Swedish policies and experiences. *Housing Studies, 25*(2), 237–256.

Andersson, Roger, & Hedman, Lina. (2015). Economic decline and residential segregation: A Swedish study with focus on Malmö. *Urban Geography.* doi:10.1080/02723638.2015.1133993.

Andersson, Roger, & Musterd, Sako. (2005). Area-based policies: A critical appraisal. *Tijdschrift voor Economische en Sociale Geografie, 96*(4), 377–389.

Batson, Christie D., & Monnat, Shannon M. (2015). Distress in the desert: Neighborhood disorder, resident satisfaction, and quality of life during the Las Vegas foreclosure crisis. *Urban Affairs Review, 51*(2), 205–238.

Bellmann, Lutz, & Gerner, Hans-Dieter (2011). Reversed roles? Wage and employment effects of the current crisis. In Herwig Immervoll, Andreas Peichl, & Konstantinos Tatsiramos (Eds.), *Who loses in the downturn? Economic crisis, employment and income distribution.* Research in Labor Economics (Vol. 32, pp. 181–206). Bingley: Emerald Group Publishing.

Belsky, Eric, & Nipson, Meg. (2010). *Long-term low income housing tax credit policy questions.* Cambridge, MA: Joint Center for Housing Studies of Harvard University.

Blank, Rebecca M. (1988). The effect of welfare and wage levels on the location decisions of female-headed households. *Journal of Urban Economics, 24*(2), 186–211.

Boelhouwer, Peter, & Priemus, Hugo. (2014). Demise of the Dutch social housing tradition: Impact of budget cuts and political changes. *Journal of Housing and the Built Environment, 29* (2), 221–235.

Bolt, Gideon, & Van Kempen, Ronald. (2011). Successful mixing? Effects of urban restructuring policies in Dutch neighbourhoods. *Tijdschrift voor Economische en Sociale Geografie, 102*(3), 361–368.

Bolt, Gideon, Van Kempen, Ronald, & Van Ham, Maarten. (2008). Minority ethnic groups in the dutch housing market: Spatial segregation, relocation dynamics and housing policy. *Urban Studies, 45*(7), 1359–1384.

Borjas, George J. (1999). Immigration and welfare magnets. *Journal of Labor Economics, 17*(4), 607–637.

Brama, Asa. (2013). The effects of neighborhood regeneration on the neighborhood hierarchy of the city: A case study in Sweden. In Maarten Van Ham, David Manley, Nick Bailey, Ludi Simpson, & Duncan MacLennan (Eds.), *Understanding neighborhood dynamics: New insights for neighborhood effects research* (pp. 111–138). Dordrecht: Springer Netherlands.

Clark, William A. V. (2013). The aftermath of the general financial crisis for the ownership society: What happened to low-income homeowners in the US? *International Journal of Housing Policy, 13*(3), 227–246.

Clark, William A. V., & Dieleman, Frans M. (1996). *Households and housing: choice and outcomes in the housing market.* New Brunswick, NJ: Centre for Urban Policy Research.

Coleman, Alice. (1985). *Utopia on trial: Vision and reality in planned housing.* London: Shipman.

Crossney, Kristen B. (2010). Is predatory mortgage lending activity spatially clustered? *The Professional Geographer, 62*(2), 153–170.

Crump, Jeff. (2002). Deconcentration by demolition: Public housing, poverty, and urban policy. *Environment and Planning D: Society and Space, 20*(5), 581–596.

Dekker, Karien, & Van Kempen, Ronald. (2004). Large housing estates in Europe: Current situation and developments. *Tijdschrift voor Economische en Sociale Geografie, 95*(5), 570–577.

Doff, Wenda, & Kleinhans, Reinout. (2011). Residential outcomes of forced relocation: Lifting a corner of the veil on neighbourhood selection. *Urban Studies, 48*(4), 661–680.

Doucet, Brian. (2014). A process of change and a changing process: Introduction to the special issue on contemporary gentrification. *Tijdschrift voor Economische en Sociale Geografie, 105* (2), 125–139.

Dreier, Peter, Bhatti, Saqib, Call, Rob, Schwartz, Alex, & Squires, George. (2014). *Underwater America: How the so-called housing 'recovery' is bypassing many American communities.* Haas Institute for a Fair and Inclusive Society. Retrieved from http://diversity.berkeley.edu/sites/default/files/HaasInsitute_UnderwaterAmerica_PUBLISH_0.pdf

Dwyer, Rachel E. (2007). Expanding homes and increasing inequalities: US housing development and the residential segregation of the affluent. *Social Problems, 54*(1), 23–46.

Ellen, Ingrid Gould, Madar, Josiah, & Weselcouch, Mary. (2014). The foreclosure crisis and community development: Exploring REO dynamics in hard-hit neighborhoods. *Housing Studies*, 1–25. (online first).

European Commission. (2010). *The European platform against poverty and social exclusion. A European framework for social and territorial cohesion* (SEC (2010), 1564 final). Brussels: Author.

Feijten, Peteke, & Van Ham, Maarten. (2009). Neighbourhood change… Reason to leave? *Urban Studies, 46*(10), 2103–2122.

Feins, Judith D., & Shroder, Mark D. (2005). Moving to opportunity: The demonstration's design and its effects on mobility. *Urban Studies, 42*(8), 1275–1299.

Fields, Desiree, & Uffer, Sabina. (2014). The financialization of rental housing: A comparative analysis of New York City and Berlin. *Urban Studies.* Advance online publication. doi:10.1177/0042098014543704

Forrest, Ray, & Hirayama, Yosuke. (2015). The financialisation of the social project: Embedded liberalism, neoliberalism and home ownership. *Urban Studies, 52*(2), 233–244.

Forrest, Ray, & Kearns, Ade. (2001). Social cohesion, social capital and the neighbourhood. *Urban Studies, 38*(12), 2125–2143.

Foster, Thomas B., & Kleit, Rachel G. (2015). The changing relationship between housing and inequality, 1980–2010. *Housing Policy Debate, 25*(1), 16–40.

Frey, William H. (2010). Race & ethnicity. In The Brookings Metropolitan Policy Program, editor, *The state of metropolitan America: On the frontlines of demographic transformation.* Washington, DC: The Brookings Institution.

Galster, George. (2001). On the nature of neighbourhood. *Urban Studies, 38*(12), 2111–2124.

Glaeser, Edward L., Resseger, Matt, & Tobio, Kristina. (2009). Inequality in cities. *Journal of Regional Science, 49*(4), 617–646.

Goetz, Edward G. (2010). Desegregation in 3D: Displacement, dispersal and development in American public housing. *Housing Studies, 25*(2), 137–158.

Goodman, Laurie, Zhu, Jun, & George, Taz. (2015). *Four million mortgage loans missing from 2009 to 2013 due to tight credit standards.* Urban Institute. Retrieved from http://www.urban.org/urban-wire/four-million-mortgage-loans-missing-2009-2013-due-tight-credit-standards

Grigsby, William, Baratz, Morton, Galster, George, & MacLennan, Duncan. (1987). The dynamic of neighborhood change and decline. *Progress in Planning, 28*(1), 1–76.

Haffner, Marietta, & Boumeester, Harry (2014). Is renting unaffordable in the Netherlands? *International Journal of Housing Policy, 14*(2), 117–140.

Hanson, Andrew, Brannon, Ike, & Hawley, Zackary. (2014). Rethinking tax benefits for home owners. *National Affairs, 19*, 40–54. Retrieved from http://www.nationalaffairs.com/publications/detail/rethinking-tax-benefits-for-home-owners

Hirschman, Albert O. (1970). *Exit, voice, and loyalty: Responses to decline in firms, organizations, and states.* Cambridge, MA: Harvard University Press.

Hoyt, Homer. (1939). *The structure and growth of residential neighborhoods in American cities.* Washington, DC: Federal Housing Administration.

Hyra, Derek, & Rugh, Jacob. (2015). The US great recession: Exploring its association with black neighborhood rise, decline and recovery. *Urban Geography* (forthcoming).

Ihlanfeldt, Keith, & Mayock, Tom. (2014). The impact of REO sales on neighborhoods and their residents. *Journal of Real Estate Finance and Economics.* Advance online publication. doi:10.1007/s11146-014-9465-0

Immergluck, Dan. (2008). From the subprime to the exotic: Excessive mortgage market risk and foreclosures. *Journal of the American Planning Association, 74*(1), 59–76.

Immergluck, Dan. (2009). The foreclosure crisis, foreclosed properties, and federal policy: Some implications for housing and community development planning. *Journal of the American Planning Association, 75*(4), 406–423.

Immergluck, Dan. (2010). The accumulation of lender-owned homes during the US mortgage crisis: Examining metropolitan REO inventories. *Housing Policy Debate, 20*(4), 619–645.

Immergluck, Dan, & Law, Jonathan. (2014). Investing in crisis: The methods, strategies, and expectations of investors in single-family foreclosed homes in distressed neighborhoods. *Housing Policy Debate, 24*(3), 568–593.

Immergluck, Dan, & Smith, Geoff. (2006). The external costs of foreclosure: The impact of single-family mortgage foreclosures on property values. *Housing Policy Debate, 17*(1), 57–79.

Immervoll, Herwig, Peichl, Andreas, & Tatsiramos, Konstantinos. (Eds.). (2011). Research in labor economics. In *Who loses in the downturn? Economic crisis, employment and income distribution* (Vol. 32). Bingley: Emerald Group Publishing.

Jivraj, Stephen. (2012). Modelling socioeconomic neighbourhood change due to internal migration in England. *Urban Studies, 49*(16), 3565–3578.

Joint Center for Housing Studies of Harvard University [JCHS]. (2015). *Improving America's housing 2015: Emerging trends in the remodeling market.* Retrieved from: http://www.jchs.harvard.edu/research/improving-americas-housing

Jones, Roderick W., & Pridemore, William Alex. (2014). A longitudinal study of the impact of home vacancy on robbery and burglary rates during the US housing crisis, 2005–2009. *Crime & Delinquency.* Advance online publication. doi:10.1177/0011128714549656

Jun, Hee-Jung (2013). Determinants of neighborhood change: A multilevel analysis. *Urban Affairs Review, 49*(3), 319–352.

Kearns, Ade, Kearns, Oliver, & Lawson, Louise (2013). Notorious places: Image, reputation, stigma. The role of newspapers in area reputations for social housing estates. *Housing Studies, 28*(4), 579–598.

Kearns, Ade, Whitley, Elise, Mason, Phil, & Bond, Lyndal. (2012). 'Living the high life'? Residential, social and psychosocial outcomes for high-rise occupants in a deprived context. *Housing Studies, 27*(1), 97–126.

Kirkness, Paul. (2014). The cités strike back: restive responses to territorial taint in the French banlieues. *Environment and Planning A, 46*(6), 1281–1296.

Kleinhans, Reinout J. (2004). Social implications of housing diversification in urban renewal: A review of recent literature. *Journal of Housing and the Built Environment, 19*(4), 367–390.

Kleinhans, Reinout J., & Bolt, Gideon. (2014). More than just fear: On the intricate interplay between perceived neighborhood disorder, collective efficacy, and action. *Journal of Urban Affairs, 36*(3), 420–446.

Kleinhans, Reinout J., & Van Ham, Maarten. (2013). Lessons learned from the largest tenure-mix operation in the world: Right to buy in the United Kingdom. *Cityscape, 15*(2), 101–117.

Kleinhans, Reinout J., & Varady, David (2011). Moving out and going down? A review of recent evidence on negative spillover effects of housing restructuring programmes in the United States and the Netherlands. *International Journal of Housing Policy, 11*(2), 155–174.

Lauria, Mickey, & Baxter, Vern. (1999). Residential mortgage foreclosure and racial transition in New Orleans. *Urban Affairs Review, 34*(6), 757–786.

Lees, Loretta. (2008). Gentrification and social mixing: Towards an inclusive urban renaissance? *Urban Studies, 45*(12), 2449–2470.

Lennartz, Christian, Arundel, Rowan, & Ronald, Richard. (2015). Younger adults and home-ownership in Europe through the global financial crisis. *Population, Space and Place.* Advance online publication. doi:10.1002/psp.1961

Lindbeck, Assar. (2006). *The welfare state. Background, achievements, problems.* Stockholm: IUI, The Research Institute of Industrial Economics.

Mallach, Alan. (2010a). REO properties, housing markets, and the shadow inventory. In Prabal Chakrabarti, Matthew Lambert, & Mary Helen Petrus (Eds.), *REO & vacant properties: Strategies for neighborhood stabilization* (pp. 13–22). Federal Reserve Banks of Boston and Cleveland and the Federal Reserve Board.

Mallach, Alan. (2010b). *Meeting the challenge of distressed property investors in America's neighborhoods.* LISC Helping Neighbors Build Communities. Retrieved from http://www. stablecommunities.org/sites/all/files/library/1228/distressed-property-investors-mallach.pdf

Martin, Anne J. (2012). *After foreclosure: The social and spatial reconstruction of everyday lives in the San Francisco Bay Area* (Doctoral dissertation). University of California Berkeley, Berkeley, CA.

Martin, Deborah G. (2003). Enacting neighborhood. *Urban Geography, 24*(5), 361–385.

Martin, Ron. (2011). The local geographies of the financial crisis: From the housing bubble to economic recession and beyond. *Journal of Economic Geography, 11*(4), 587–618.

Mayer, Christopher J., & Pence, Karen. (2008). *Subprime mortgages: What, where, and to whom?* Cambridge, MA: The National Bureau of Economic Research.

McPherson, Miller, Smith-Lovin, Lynn, & Cook, James M. (2001). Birds of a feather: Homophily in social networks. *Annual Review of Sociology, 27*, 415–444.

Meen, Geoffrey, Nygaard, Christian, & Meen, Julia. (2013). The causes of long-term neighbor-hood change. In Maarten Van Ham, David Manley, Nick Bailey, Ludi Simpson, & Duncan MacLennan (Eds.), *Understanding neighborhood dynamics: new insights for neighborhood effects research* (pp. 43–62). Dordrecht: Springer Netherlands.

Morris, David, & Hess, Karl. (1975). *Neighborhood power: The new localism.* Boston, MA: Beacon Press.

Mulder, Clara H., & Hooimeijer, Pieter. (1999). Residential relocations in the life course. In Leo Van Wissen & Pearl Dykstra (Eds.), *Population issues: An interdisciplinary focus* (pp. 159–186). The Hague: NIDI.

Musterd, Sako, & Ostendorf, Wim. (2005). Social exclusion, segregation, and neighborhood effects. In Yuri Kazepov (Ed.), *Cities of Europe: Changing contexts, local arrangements and the challenge to urban cohesion* (pp. 170–189). Oxford: Blackwell.

Newman, Kathe. (2009). Post-industrial widgets: Capital flows and the production of the urban. *International Journal of Urban and Regional Research, 33*(2), 314–331.

Newman, Oscar. (1972). *Defensible space.* New York, NY: Macmillan.

OECD. (2013). *Crisis squeezes income and puts pressure on inequality and poverty.* Retrieved from: http://www.oecd.org/els/soc/OECD2013-Inequality-and-Poverty-8p.pdf

OECD. (2014). *OECD economic outlook 2014.* Retrieved from: http://www.oecd.org/els/oecd-employment-outlook-19991266.htm

Ojeda, Raul H. (2009). *The continuing home foreclosure tsunami: Disproportionate impacts on black and Latino communities* (WCVI White Paper). San Antonio, TX: William C. Velasquez Institute.

Peck, Jamie. (2012). Austerity urbanism. *City, 16*(6), 626–655.

Permentier, Matthieu, Van Ham, Maarten, & Bolt, Gideon. (2007). Behavioural responses to neighbourhood reputations. *Journal of Housing and the Built Environment, 22*(2), 199–213.

Pfeiffer, Deirdre, & Molina, Emily T. (2013). The trajectory of REOs in Southern California Latino neighborhoods: An uneven geography of recovery. *Housing Policy Debate, 23*(1), 81–109.

Phillips, Deborah, & Harrison, Malcolm. (2010). Constructing an integrated society: Historical lessons for tackling black and minority ethnic housing segregation in Britain. *Housing Studies, 25*(2), 221–235.

Posthumus, Hanneke, Bolt, Gideon, & Van Kempen, Ronald. (2013). Why do displaced residents move to socioeconomically disadvantaged neighbourhoods? *Housing Studies, 28*(2), 272–293.

Power, Anne. (1997). *Estates on the edge: The social construction of mass housing in Northern Europe.* London: Macmillan.

Prak, Niels L., & Priemus, Hugo. (1986). A model for the analysis of the decline of postwar housing. *International Journal of Urban and Regional Research, 10*(1), 1–7.

Priemus, Hugo, & Whitehead, Christine. (2014). Interactions between the financial crisis and national housing markets. *Journal of Housing and the Built Environment, 29*(2), 193–200.

Puno, Norene J., & Thomas, Mark. (2010). *Interrogating the new economy: Restructuring work in the 21st century.* Toronto, ON: University of Toronto Press.

Putnam, Robert D. (2007). E pluribus unum: Diversity and community in the twenty-first century. The 2006 Johan Skytte prize lecture. *Scandinavian Political Studies, 30*(2), 137–174.

Raco, Mike, & Tasan-Kok, Tuna. (2009). Competitiveness, cohesion, and the credit crunch: Reflections on the sustainability of urban policy. In Katrien De Boyser, Caroline Dewilde, Danielle Dierckx, & Jurgen Friedrichs (Eds.), *Between the social and the spatial: Exploring the multiple dimensions of poverty and social exclusion* (pp. 183–195). Farnham: Ashgate.

Robertson, Douglas, McIntosh, Ian, & Smyth, James (2010). Neighbourhood identity: The path dependency of class and place. *Housing, Theory and Society, 27*(3), 258–273.

Rohe, William M., Van Zandt, Shannon, & McCarthy, George. (2002). Home ownership and access to opportunity. *Housing Studies, 17*(1), 51–61.

Ross, Catherine E., Mirowsky, John, & Pribesh, Shana. (2001). Powerlessness and the amplification of threat: Neighborhood disadvantage, disorder, and mistrust. *American Sociological Review, 66*(4), 568–591.

Rugh, Jacob S., & Massey, Douglas S. (2010). Racial segregation and the American foreclosure crisis. *American Sociological Review, 75*(5), 629–651.

Sampson, Robert J. (2009). Disparity and diversity in the contemporary city: Social (dis)order revisited. *The British Journal of Sociology, 60*(1), 1–31.

Scanlon, Kathleen, & Elsinga, Marja. (2014). Policy changes affecting housing and mortgage markets: How governments in the UK and the Netherlands responded to the GFC. *Journal of Housing and the Built Environment, 29*(2), 335–360.

Schelkle, Waltraud. (2012). A crisis of what? Mortgage credit markets and the social policy of promoting homeownership in the United States and in Europe. *Politics & Society, 40*(1), 59–80.

Schloemer, Ellen, Wei, Li, Ernst, Keith, & Keest, Kathleen. (2006). *Losing ground: Foreclosures in the subprime market and their cost to homeowners.* Center for Responsible Lending. Retrieved from http://www.responsiblelending.org/mortgage-lending/research-analysis/foreclosure-paper-report-2-17.pdf

Schneider, Martin, & Wagner, Karin. (2015). Housing markets in Austria, Germany and Switzerland. In Ernest Gnan, Doris Ritzberger-Grünwald, Helene Schuberth, & Martin Summer (Eds.), *Monetary policy and the economy: Quarterly review of economic policy* (pp. 42–58). Vienna: Oesterreichische Nationalbank.

Schwartz, Alex. (2011). The credit crunch and subsidized low-income housing: The UK and US experience compared. *Journal of Housing and the Built Environment, 26*(3), 353–374.

Shaw, Clifford R., & McKay, Henry D. (1942). *Juvenile delinquency and urban areas.* Chicago, IL: University of Chicago Press.

Shierholz, Heidi. (2014). *The truth behind today's long-term unemployment crisis and solutions to address it.* Economic Policy Institute. Retrieved from http://www.epi.org/publication/truth-todays-long-term-unemployment-crisis/

Skifter Andersen, Hans (1999). Housing rehabilitation and urban renewal in Europe: A cross-national analysis of problems and policies. In Hans Skifter Andersen & Philip Leather (Eds.), *Housing renewal in Europe* (pp. 241–277). Bristol: The Policy Press.

Strom, Elizabeth, & Reader, Steven. (2013). Rethinking foreclosure dynamics in a Sunbelt city: What parcel-level mortgage data can teach us about subprime lending and foreclosures. *Housing Policy Debate, 23*(1), 59–79.

Swank, Duane. (1998). Funding the welfare state: Globalization and the taxation of business in advanced market economies. *Political Studies, 46*(4), 671–692.

Tammaru, Tiit, Marcińczak, Szimon, Van Ham, Maarten, & Musterd, Sako (Eds.). (2015). *Socio-economic segregation in European capital cities: East meets west.* Oxford: Routledge.

Teernstra, Annalies. (2014). Neighbourhood change, mobility and incumbent processes: Exploring income developments of in-migrants, out-migrants and non-migrants of neighbourhoods. *Urban Studies, 51*(5), 978–999.

Thomas, Hannah. (2013). The financial crisis hits home: Foreclosures and asset exhaustion in Boston. *Housing Policy Debate, 23*(4), 738–764.

Tsenkova, Sasha, & Turner, Bengt. (2004). The future of social housing in Eastern Europe: Reforms in Latvia and Ukraine. *European Journal of Housing Policy, 4*(2), 133–149.

Tunstall, Rebecca. (2015). Are neighborhoods dynamic or are they slothful? The limited prevalence and extent of change in neighborhood socio-economic status, and its implications for regeneration policy. *Urban Geography* (forthcoming).

Van Beckhoven, Ellen, Bolt, Gideon, & Van Kempen, Ronald. (2009). Theories of neighborhood change and neighborhood decline: Their significance for post-WWII large housing estates. In Sako Musterd, Ronald Van Kempen, & Rob Rowlands (Eds.), *Mass housing in Europe: Multiple faces of development, change and response* (pp. 20–50). Basingstoke: Palgrave MacMillan.

Van Der Heijden, Harry, Dol, Kees, & Oxley, Michael (2011). Western European housing systems and the impact of the international financial crisis. *Journal of Housing and the Built Environment, 26*(3), 295–313.

Van Eijk, Gwen. (2010). *Unequal networks: Spatial segregation, relationships and inequality in the city.* Delft: IOS Press.

Van Gent, Wouter, Musterd, Sako, & Ostendorf, Wim. (2009). Bridging the social divide? reflections on current Dutch neighbourhood policy. *Journal of Housing and the Built Environment, 24*(3), 357–368.

Van Ham, Maarten, & Manley, D. (2012). Neighbourhood effects research at a crossroads. Ten challenges for future research. *Environment and Planning A, 44*(12), 2787–2793.

Van Kempen, Ronald, & Priemus, Hugo. (2002). Revolution in social housing in the Netherlands: Possible effects of new housing policies. *Urban Studies, 39*(2), 237–253.

Wachter, Susan. (2015). The housing and credit bubbles in the United States and Europe: A comparison. *Journal of Money, Credit and Banking, 47*(S1), 37–42.

Wacquant, Loïc. (2008). *Urban outcasts: A comparative sociology of advanced marginality.* Cambridge, UK: Polity Press.

Warren, Donald. (1981). *Helping networks: How people cope with problems in the urban community.* South Bend, IN: Notre Dame University Press.

Watson, Matthew. (2009). Planning for a future of asset-based welfare? New Labour, financialized economic agency and the housing market. *Planning Practice & Research, 24*(1), 41–56.

Whitehead, Christine, Scanlon, Kathleen, & Lunde, Jens. (2014). *The impact of the financial crisis on European housing systems: A review.* Stockholm: Swedish Institute for European Policy Studies.

Wilson, James Q, & Kelling, George L. (1982). Broken windows. *Atlantic Monthly, 249*(3), 29–38.

Zwiers, Merle D., & Koster, Ferry. (2015). The local structure of the welfare state: Uneven effects of social spending on poverty within countries. *Urban Studies, 52*(1), 87–102.

Reclaiming neighborhood from the inside out: regionalism, globalization, and critical community development

Kathe Newman[a] and Edward Goetz[b]

[a]Urban Planning and Policy Development, Rutgers University, New Brunswick, NJ, USA; [b]Humphrey School of Public Affairs, University of Minnesota, Minneapolis, MN, USA

ABSTRACT

In this article, we argue the need for a critical turn in community development practice and research in the face of two scalar tensions in the existing academic literature and US community development policy. The first tension is the perceived ineffectiveness of neighborhood-based community development in the context of globalization despite the increasing interrelatedness of neighborhoods and globalization. The second tension emerges in a growing body of academic literature and policy action that privileges the region as the place from which to understand urban decline and to address issues that have historically been the concern of community development. These two tensions dominate community development discussions and often undermine community development politics and policy by contesting its relevance for multiscalar processes. Instead, we argue that neighborhoods are more important than ever because it is from the place of the neighborhood that it becomes possible to understand the multiscalar (global and local) processes that shape it.

Introduction

In this article, we argue the need for a critical turn in community development practice and research in the face of two scalar tensions in the existing academic literature and US community development policy. The first tension is the perceived ineffectiveness of neighborhood-based community development in the context of globalization. The second tension privileges the region as the place from which to understand urban decline and to address issues that have historically been the concern of community development. These two tensions dominate community development discussions and often undermine community development politics and policy by contesting its relevance for multiscalar processes. We argue that both of these challenges, the global and the regional, can be addressed by community development adopting as one of its objectives, affecting what Pierce, Martin, and Murphy (2011) call a politics of place—defined as a politics of the multiscalar processes that shape places (DeFilippis, 1999; Lake, 1994; Martin, 2004). While many processes shape the politics of place, we use the example of the financialization of the economy and the recent housing foreclosure crisis to illustrate the broader concepts and to highlight the connections between globalization and urban neighborhoods.

The article is organized in two main sections. In the first section, we discuss the tension between local and global. Community development is often seen as a place-based effort with little impact on global processes. Such an understanding of community development consigns local efforts to a strategic position somewhere between impotence and irrelevance and ignores the actually existing relationships between neighborhoods and the broader world. The urban studies literature on relational place-making contains, however, two related yet distinct discussions about actually existing relationships between neighborhoods and globalization, which, when brought together, can inform a critical approach to community development policy and practice. The broader discussion is about the politics of place-making which highlights how places are shaped not only by what happens within but by a bundle of actions, actors, and processes that come together to shape it regardless of where they are located. Pierce et al. (2011) have synthesized these literatures as processes of relational place-making. A related discussion is about the urbanization of capital and the financialization of the economy which has drawn urban neighborhoods and globalizing processes ever closer together despite their geographic separation. Given the recent foreclosure and global economic crisis, and how it has affected many urban neighborhoods, we discuss this literature to illuminate some of the processes that shape relational place-making.

The relational place-making approach greatly informs community development literature and practice since it expands the horizon for place politics and pushes community development scholars and practitioners to identify the processes and politics that shape urban neighborhoods. In contrast with a community development approach that looks at a jurisdictionally defined place surrounded by thick borders, the approach adopted in the place-making literature starts with a place and adopts fluid borders that better approximate the flow of actors and processes through place. The politics of place-making is the politics about the decisions that shape the place of study regardless of where those decisions take place (Lake, 1994). For community development, this means looking at what foreclosure means in urban neighborhoods and why it happened which necessitates an exploration of processes and actors outside of neighborhoods. For example, the collateral for many subprime home loans were homes in inner-city US neighborhoods; investors and financial institutions were located all over the world; and a variety of local, national, and global regulatory bodies established the rules and markets that made the financial transactions possible. To understand urban change then means exploring how multiscalar processes shape neighborhoods. After discussing these literatures, and an example of the actual experiences of the financializing economy and its relationship to urbanization, we argue the need for a critical approach to community development that reconsiders neighborhoods in a globalizing world and reconceptualizes community development politics as a politics of place-making which we define as contesting the processes through which places are made.

In the second section, we consider the tension in literature and policy narratives between neighborhoods and region. Metropolitan or regional approaches to understanding urban decline and renewal diminish the importance of neighborhood-based community development efforts in the United States as a policy objective with a discourse that marginalizes local renewal in favor of mobility strategies targeting a metropolitan level "geography of opportunity"; in the United Kingdom and some other European countries, similar policies focus on social mixing. These regional mobility efforts are largely aimed at moving people around to access opportunity, and are framed as initiatives to deconcentrate poverty. These regionalist initiatives have focused on narrow explanations of urban poverty and economic decline and have been justified by the alleged failure of community development efforts to revive urban neighborhoods (Goetz, 2003; Goetz & Chapple, 2010). While regionalists were right to look beyond the boundaries of the neighborhood for the processes that shape neighborhood change, they looked to another level (the metropolitan area or region) rather than to the processes and politics that shape urban neighborhoods, something Harvey (1987/ 1994) cautions against. Arguably the processes that created metropolitan uneven development were at one time important sources of urban place-making, but the way in which regionalists approach regional strategies does not address the politics around uneven development. Instead, community development should take on, as one of its objectives, engagement with a politics of place-making that directly addresses globalization and other processes from the perspective of neighborhoods.

In the conclusion we suggest a critical approach to community development literature and practice that illuminates the politics of relational place-making (Pierce et al., 2011). We argue for a community development strategy that strengthens democracy within communities and builds organizations that work to recast the politics of multiscalar processes that shape neighborhoods and communities. Such a strategy, we maintain, can address the dynamics of financialization on the one hand, making community development relevant and effective in addressing globalized investment practices, while simultaneously providing the basis for a response to regional inequities that go beyond the limited mobility solutions of regionalism and focus on the immediate needs of people in urban neighborhoods.

The neighborhood and the world

The first scalar mismatch is the tension in community development literature and practice between neighborhood and globalization. Neighborhood and globalization are often viewed as distinct spheres in which globalization affects neighborhoods, and neighborhood development efforts are perceived as having little effect on globalizing processes. But a body of urban studies literature illuminates the interrelationships between neighborhoods and globalization, which, we argue, can help frame a more critical approach to community development literature and practice. There are two distinct discussions within this literature: literature that broadly (1) views neighborhood change as the product of a multiscalar politics of place-making; or (2) views urban change as shaped through the urbanization of capital and through globalization and financialization processes.

Relational place-making and multiscalar processes

The literature on place-making is a critical effort to operationalize Lefebvre's ideas about unpacking the processes that shape urban change by looking at urbanization and the processes that shape it. Pierce et al. (2011, p. 58) describe relational place-making as individuals making places "by referencing and (re)configuring the many simultaneous places that they participate in; these places-bundles are socially negotiated, constantly changing and contingent." As with the previous discussion of capital flows, this approach helps to illuminate how places are interrelated.

> Our intervention aims to help geographers by providing an analytical approach … to consider the interconnections and co-constituencies among place, networks and politics by identifying specific conflicts and the places they produce, the dimensions of place-framing evident, and the multiply-positioned actors and places/bundles inherent in and underlying the conflicts. (Pierce et al., 2011, p. 67)

In many ways, the politics of place is one object of community development and it means identifying and engaging in the decisions that shape places wherever those decisions take place (Cox, 1998; DeFilippis, 1999; Martin, 2004). The literature on place-making stresses that place-making "…draws on elements at multiple scales; the connectivities and contingencies that shape a place are not at all limited to the scale of that place" (Pierce et al., 2011, p. 59). But the actors and processes that shape places come together in place and time as bundles. Given that these processes flow across the globe, what happens in one place may well shape what happens on the other side of the world. Financial regulations in the United Kingdom, for example, could well affect the cost and availability of capital in the United States, which could shape urban places. Global regulations on bank capitalization and efforts to reduce speculation could affect local communities and local communities could work with communities across the globe to construct rules that shape their respective places. Thinking about place-making, as shaped by place interrelationships, opens up community development to innovative approaches to explore the politics of place-making. The last decade has seen the emergence of an array of new community development efforts including the Right to the City Coalition, which brings people together to explore the processes that shape their communities. We turn now to the second set of discussions to consider one set of processes that draw urban neighborhoods and globalizing processes together.

Globalization and financialization

In the wake of deindustrialization, scholars and practitioners have sought to understand the political economy of the post-industrial city. While some saw cities fading away in a globalizing world dominated by information and population flows, Sassen (2005) argued that the command and control functions of a global economy would locate in a handful of global cities making them important sites for development. The demand for highly skilled labor would lead to increased inequality in these cities and a growth in informalization would make it possible for the less skilled to compete (Sassen, 2005). Harvey (1987/1994, p. 381) too believed in the importance of place but cautioned against seeing places and scales as distinct: "It is invidious to regard places, communities, cities, regions or even nations as 'things in themselves' at a time when the global flexibility of capitalism is greater than ever." Harvey (1987/1994, p. 381) added "To follow that line of thinking is to be increasingly rather than less vulnerable in aggregate to the extraordinary centralized power of flexible accumulation." For Harvey, it was impossible to understand the city in the world without understanding the ways in which cities were interrelated. Looking only at the local made it all too easy to miss the real sources of urban change, which in his mind (1987/1994, p. 382) "define a politics of adaptation and submission." In the context of the recent foreclosure and global financial crisis, looking only at the local or even the region misses the broader changes of globalization and the financialization of the economy. It misses the speed and fluidity with which capital accumulation logics dominate borrowing and lending and prioritizes profit over how communities used their savings in the past. Ignoring these processes and the decisions that shape them misses vital processes of political decision-making that shape urban neighborhoods.

While Sassen (2001) was right about cities like New York, London, and Shanghai, in that the command and control functions centered there and economic change brought higher skilled and paid workers to the city center, this was only part of the story. The literature on the financialization of the economy (the increasing role of finance in generating economic profits) suggests even more direct relationships between neighborhoods and globalizing processes. While global capital is often viewed as fluid and placeless, Harvey views flexible accumulation as circulating the globe, seeking yield through a spatial fix (Harvey, 1989). Martin (2011, p. 591) explains the implication for the relationship between urbanization and globalization: "A major consequence of globalization has been to create new relational and functional monetary spaces that are simultaneously geographically compressed and geographically stretched." Urban communities and the world are closer together even though they are far apart.

The commodification of housing and finance has woven neighborhoods and home ever more tightly into global capital markets and conversely has woven the logics of capital markets ever more tightly into homes and neighborhoods (Fox Gotham, 2006). Langley (2006) shows how home mortgage securitization, the process of issuing securities based on home mortgage principal and interest income streams, has drawn neighborhoods and the logics of capital accumulation together. Modern global privatized banking systems have largely replaced the traditional banking systems that once linked, or intermediated, local communities of savers and borrowers. Borrowing money to purchase housing today often means accessing capital on global markets and involves a variety of financial institutions and a web of rules that shape those institutions, transactions, and markets (Christophers, 2014; Langley, 2006; Leyshon & Thrift, 2007; Moreno, 2014; Wyly et al., 2006).

While an approach following Sassen sees residential demand and urban change as the result of new workers and companies seeking space and driving up prices, an approach following Langley and Christophers sees the place of neighborhoods and homes shaped in part by the logics of global finance. The story of neighborhood change is then about the supply of capital as much as it is about its demand (Smith, 1996). And this then sets fundamentally new challenges for community development especially because these systems are complex, opaque, privatized, global, and often fast moving. While many urban borrowers struggled to access capital in the past, changes within the global financial system made borrowing possible for many more people, which some initially celebrated as the result of decades of community development organizing. In the United States, innovations in securitization enabled financial institutions to originate loans that were not possible prior to this period of financial innovation and as interest rates rose and prime borrowers dropped out, the system sought more borrowers and created more flexible loan products and loan terms to appeal to them

(McCoy, Pavlov, & Wachter, 2009). Financial institutions loosened underwriting standards in 2005 and 2006 and originated more nontraditional loans such as interest only and pay option loans that made it possible for an even broader array of borrowers to access loans and meant that borrowers were not paying part or all of the principal on their loans with the result that loan to value ratios could exceed 100%. Borrowers could quickly owe more than their homes were worth (Demyank & Gopalan, 2007: online; McCoy, Pavlov, Wachter, 2009). With interest rates at the lowest point in 2003, since the Federal Reserve provided online historical data, many borrowers refinanced. The number of 1–4 family home refinance mortgage organizations in the United States increased from 2,435,420 in 2000 to 7,889,186 in 2001 and exceeded 10,000,000 in 2002 and 15,000,000 in 2003 before dropping to 7,583,928 in 2004 as interest rates rose. The number of 1–4 family home purchase loans increased from 4,938,809 in 2001 to a peak of 7,382,012 in 2005 (Avery, Bhutta, Brevoort, & Canner, 2012; Federal Reserve Bank of St. Louis, 2014). Meanwhile, private agencies increased their share of mortgage-backed security issuances competing with the historically more conservative government sponsored enterprises (GSE), which operate with the implicit backing of the US government. Non-GSE mortgage entities issued a considerably larger share of mortgage backed securities in 2004, 2005 and 2006 than they had before (Goodman, 2015).

The effect of low-cost accessible credit was felt directly in urban places that witnessed rising housing prices and new development with some cities declaring that they were finally back from the fiscal abyss of the 1970s. The drawing of global capital into urban neighborhoods through individual borrowers had the effect of capital spreading out through communities, weaving neighborhoods into the global economy house-by-house, apartment-by-apartment. In the 1970s and 1980s, Neil Smith described the gentrification frontier as the edge of changing neighborhoods where developers would purchase buildings pushing ever farther outward but still close enough to already higher income areas to reduce their risk (Smith, 1996). Mortgage securitization made it possible to link global capital to some of the most disinvested communities in the country, lot-by-lot, unit-by-unit. The places where banks had feared to lend became places of expansion for the global financialized economy in the post-industrial city.

Historically, the availability and cost of capital has dramatically shaped urbanization leading to a politics of community development reinvestment in the 1970s and two major pieces of federal legislation in the United States—the Home Mortgage Disclosure Act, which requires gathering information on mortgage lending at the census tract level, and the Community Reinvestment Act, which required lenders to make loans in the communities where they take deposits. While redlining during the twentieth century made it difficult for some people to buy or renovate homes and contributed to urban decline, the availability of low-cost loosely underwritten capital during the middle part of the first decade of the twenty-first century contributed to redevelopment, home price increases, and in many places, foreclosures.

Looking at these processes from the place of the local foregrounds the vitality and significance of everyday life and how they are shaped by their interaction with these global processes. The literature on financialization and urbanization argues that securitization links the logic of global capital accumulation directly to neighborhoods through the place of home. While many borrowers accessed lower-cost capital through these processes, purchased homes, and increased their equity, borrowers in and residents of many inner-city communities had vastly different experiences that involved displacement and foreclosure. Some found themselves living on blocks filled with vacant homes and others found themselves in increasingly expensive neighborhoods. That the financialization of the economy has drawn global capital markets and community ever closer together suggests its importance for everyday life (DeFilippis, 2009; Newman & Teresa, 2013). Neighborhoods then remain at the heart of community development policy and practice but part of the community development agenda means understanding and engaging with the politics of place, the processes that shape neighborhoods.

The academic literature on neighborhoods and globalization and especially recent work on interrelational place-making opens the door to better illuminating the processes of change in urban neighborhoods which can help inform a twenty-first century approach to community development. But the second scale mismatch, regional–local, has also consumed a considerable amount of academic and policy attention and it has diverted attention from neighborhoods to the place of the region. We turn our attention to that scale mismatch now and consider how the idea behind us can move community development forward without losing neighborhood.

The neighborhood and the metropolitan region

The second scale mismatch that faces community development is the regional–local one. This tension presents a few challenges. Regionalism we define as a political and policy approach that locates the problems of central city neighborhoods, as well as the solutions to those problems, in the relationship of those neighborhoods to larger, metropolitan-level economic, social, and political dynamics (see Cisneros, 1995; Imbroscio, 2010; Pastor, Peter Dreier, Grigsby, & Marta, 2000; Swanstrom, 1995). In other words, regionalists reject the neighborhood in favor of the region. As a policy approach, regionalism has eclipsed community development over the past 20 years in the United States, and in the process has marginalized practical community-based efforts. One challenge facing community development is to confront its own waning position and the discourse that relegates neighborhood-based efforts to the category of "tried and failed" approaches that are often related to discussions about further concentrating poverty and curtailing urban opportunity. Reviving community development will require employing a community-based relational (not locally bounded) politics. The place-making literature suggests the importance of beginning with places but not constraining understandings of place-making as only those things that unfold within places. This approach suggests a challenge to the dominant regionalist paradigm that characterizes much American urban work at the moment, but it also requires a greater examination of how responses to nonlocal processes should inform and become part of community-based revitalization. This introduces the second challenge presented by the regional–local scale mismatch— theorizing a more contextualized community development approach. The movement needs to move beyond an inward orientation that identifies problems and solutions both as internally determined.

The triumph of regionalism

American urban policy dogma since the early 1990s has held that decades of place-based revitalization policy, from urban renewal through the expansion of the community development movement in the 1980s, has not effectively reversed the decline of central city neighborhoods in American cities (Orfield, 1997; Pastor, Brenner, & Matsuoka, 2009; Rusk, 1993). There are three legs to this stool: (1) the accusation of failure of community development, (2) the "discovery" of concentrated poverty and extreme marginalization in core (i.e., inner-city) neighborhoods of American urban areas, and (3) the articulation of a new set of policy theories to frame urban policy. The first of these is the accusation of failure. Lehmann (1994) in a now famous *New York Times Magazine* story, proclaimed the failure of place-based urban policy, documenting the millions of dollars in public spending that had been channeled into declining neighborhoods, and the apparent lack of positive impact it had produced. Leading regionalist thinkers echo this belief about the limitations of neighborhood-centered, place-based efforts. Pastor et al. (2009, p. 9) see community development as "the equivalent of swimming against a raging stream" of "policies and contexts" that are pushing investment outward (see also Orfield, 1997). Rusk (1999) likens it to helping people "up a down escalator." These metaphors evoke the sense of a larger system of inequality that community development efforts ignore or are unsuited to challenge, while they focus on more limited and proximate concerns that might increase the upward mobility of a fortunate few. The constrained efforts of community development, bounded by defined place limits and confined to a small set of policy areas (primarily housing and economic development) are, according to the regionalist narrative, incapable of addressing the broader metropolitan dynamics shaping and maintaining neighborhood poverty and disadvantage.

The second source of community development's crisis of legitimacy is the "discovery" of concentrated poverty and the growing pathologizing of central city neighborhoods. The work of Wilson and others in naming and measuring concentrated poverty suggested, in fact, that conditions in American cities were demonstrably worse as the 1990s began than they had been in the 1960s before the community development movement began. Drugs, high rates of violent crime, and the specter of a dangerous class of alienated and marginalized surplus population (largely young, male, and Black) dominated the discourse of popular media, academia, and policy circles during the late 1980s and 1990s (Macek, 2006). The emergence of urban "no-go" zones, portrayed in anxious news stories of crime, gangs, and violence in cities, replete with exaggerated misinformation about crack and cocaine (see Reeves & Campbell, 1994), fed the growing sense that previous community development policy had been ineffective.

In the regionalist narrative, the prevailing image of disadvantaged neighborhoods became one in which the poor and people of color are trapped by impersonal and, importantly, *nonlocal* policies such as federal homeownership programs that induced white flight from central cities, exclusionary zoning in suburbs that prevented affordable housing options for the poor, and urban renewal, public housing, and interstate highway development that conspired to concentrate low-cost housing in core neighborhoods (Dreier, Mollenkopf, & Swanstrom, 2001; Sharkey, 2013; Wilson, 1987). The middle class and productive economic activity, in this analysis, had been leaving core neighborhoods for years, leaving behind a more uniformly poor and marginalized population. In the light of this demographic understanding of neighborhood distress, the actions of community developers themselves became suspect in that they were said to anchor the poor in core and disadvantaged neighborhoods (McGeary, 1990).

Third, neighborhood-based community development could be marginalized because there was a ready substitute—the regionalism project. As a body of thought it incorporates the notion of community development's flaws and shortcomings, as described earlier. In fact, regionalism critiques community development while sharing many of the overriding objectives of reducing inequities and improving life chances of those living in disadvantaged neighborhoods (Dreier et al., 2001; Pastor et al., 2009). Regionalism incorporates a theory of place-based effects that focuses on the debilitating nature of neighborhood environment as an explanation for multigenerational urban poverty (Jencks & Mayer, 1990; Massey & Denton, 1993).

The political success of the regionalist analysis was almost total. Regionalists reacted to Reagan-era federalism reforms that devolved redistributive responsibility to the local and state level. Foundations, policy think tanks, and the federal government embraced the notion of the pathological neighborhood, the place that resisted meaningful internal improvement and required instead a new approach that focused on outside factors that determined regional patterns of segregation and investment. Foundations began to support regional organizing efforts and state legislatures became the target for regional reform efforts. Yet, despite an analysis of the problem that included the identification of racism and racial segregation, the exclusionary acts of predominantly White and affluent suburban areas, and the regional dynamics of economic development, the set of regionalist policy initiatives adopted to date has been quite narrow in scope and scale. This "primitive regionalism," in fact, bears little resemblance to the "outside game" that Rusk (1999) advocated. Little headway has been made, for example, in creating or expanding regional governance that would theoretically overcome the parochialism of fragmented governments. Little has changed about the authority of suburbs to limit affordable housing development within their jurisdictions. There is no more tax-sharing activity taking place in metropolitan areas now than there was 30 years ago. Private sector investment patterns and much public sector investment continue to follow logics that have little to do with redressing inequities across metropolitan areas.

Primitive regionalism instead operates through a relatively small number of initiatives that mainly function to redistribute people across the metropolitan landscape, and serves mostly to change core neighborhoods in important ways. So-called mobility programs, for example, from the influential Gautreaux program in Chicago to the national Moving to Opportunity (MTO) program, serve to allow a lucky few low-income families to move out of high-poverty neighborhoods using housing vouchers. In addition to inviting the poor to move, programs of redevelopment like HOPE VI and the Choice Neighborhoods Initiative (CNI) compel the poor to move by funding the demolition of low-cost housing and wholesale redevelopment, forcing displacement and relocation. To these efforts are added others that attempt to attract middle-income households to core neighborhoods that are no longer characterized by concentrations of subsidized, affordable housing.

Thus, primitive regionalism has been little more than the deconcentration of poverty through the manipulation of assisted housing policy. In this perspective, neighborhood revitalization is seen, first and foremost, as the subtraction of poor people from neighborhoods and their replacement by more affluent ones. There are corollary theories and objectives being served such as the notion that architecture (specifically, the modernist architecture of much public housing) helped to produce ghettos and that the porches of new urbanism will be an effective antidote (Epp, 1996). But, the lead factor in regionalism as it has been practiced in the past 20 years has been displacement of lower-income residents and their replacement by households with greater means. As that is accomplished, the state and private sector have combined to reintroduce millions of dollars for redevelopment efforts to revitalize neighborhoods and to provide more fertile ground for market-based investment. Primitive regionalism has therefore become the context for forced displacement of the poor, gentrification, and the continued marginalization of community-led, place-based neighborhood efforts.

There is, in the regionalist approach, an assumption of desired mobility on the part of the poor, a desire they presumably cannot act on because of the traps represented by spatially concentrated services for the poor and the exclusionary tactics of other communities (Imbroscio, 2008; 2012). This is important because it is another way of expressing the notion that core neighborhoods are not desirable, or not only cannot be saved through place-based, community efforts, but ought not to be saved and should be reshaped through a dramatic replacement of low-income people with high-income people. It implies that the solutions to inner-city poverty lie in mobility. While mobility may benefit some, the approach distracts from efforts to understand the politics of place-making and to develop an engaged response to shape those efforts to improve the lives of people in their communities.

Resolving the local–global and local–regional tensions

Equally important for addressing both the global–local and regional–local scale mismatches is recasting the idea of the bounded localism of past community development efforts. The early history of community development in the United States is one of organizing against public and private power brokers. The community action initiatives of the war on poverty gave rise to vigorous analyses of race and power differentials in urban areas, and were focused on a reallocation and redistribution of government services and decision-making power. In the 1970s, community development activists identified and targeted lending institutions for their pattern of systematic and discriminatory disinvestment. The neighborhood movement of the time led to significant national legislation reforming the lending industry and making lending practices more transparent (Rosenbloom, 1981).

As political backlash to community action emerged and grew, however, and as the constraints of municipal budgets and federal programs became more apparent, community development shifted to a more self-help strategy. Initially, this strategy, too, was nested within a radical theory of replacing outside institutions with local ownership and control. But, as the movement began to professionalize and become more pragmatic, it moved away from organizing and confrontation, and became divorced from a political understanding of neighborhood issues. DeFilippis (2004) argues that the movement has progressed further down this path in the past 20 years into what he characterizes as "neo-liberal communitarianism." This identifies community development as a nonconfrontational movement aimed at making market processes and government investment practices (the basis of which go largely unquestioned by the movement) work for their own communities, and that these communities and their residents "have a shared set of interests with the larger society."

The criticism of the regionalists is relevant here when they argue that too much neighborhood-based work does not engage with the broader economic and political processes that help to produce local conditions. Yet regionalist policy initiatives largely miss the broader political economic processes as well, and thus do little to address the deficiencies of bounded localism. MTO and Gautreaux, for example, ignore Rusk's downward escalator just as completely as any community development initiative that regionalists critique. Further, primitive regionalism is problematic even on its own terms in that it elides the challenges involved in moving people from one location to another, underestimating or neglecting the inadequacy of transportation and social service infrastructure in suburban areas. The embedded assumption of primitive regionalism, that location will solve all problems, all but ignores broader forces in favor of reshuffling the demographic deck of metropolitan areas and it completely misses the increasing relationships between neighborhoods and global processes.

Addressing the scale mismatch between local efforts and regional dynamics, for example, requires adopting some of the regionalist perspective while rejecting those elements that problematize core communities. It also requires rejecting an overly simple, binary reality in which some communities lack opportunity/amenities and others do not (or alternatively, some neighborhoods are problematic while others are not). Most promising in this respect are the comprehensive "equity analyses" emerging in metropolitan areas across the country. Efforts to map regional inequities in the past 10 years examine opportunity as a multidimensional phenomenon, encompassing access to jobs, healthy living conditions, transit, affordable housing, and public services (see, e.g., Sadler, Wampler, Wood, Barry, & Wirfs-Brock, 2012). A more critical approach would start with the neighborhoods and explore the politics of conflict wherever those were located. Instead of predefining the place of the metropolitan area, a critical approach to community development could remove the regional barriers to explore the world.

Analyses that acknowledge the multidimensionality of opportunity recognize the different forms of opportunity that exist across neighborhoods, and move beyond the dualistic primitive regionalist framework that pathologizes core neighborhoods. Examples of this include the "regional equity analyses" produced in Los Angeles, Atlanta, Denver, and Portland (e.g., Sadler et al., 2012). Such analyses produce (either literally or in-effect) a "place–opportunity" matrix in which different neighborhoods or sub-areas of the region are assessed by each of the dimensions of opportunity that are studied. The place–opportunity matrix in turn provides a framework for effective community development efforts by highlighting the need for effective local action to preserve and enhance the amenities and services within core neighborhoods, in addition to challenging exclusionary tactics of suburban communities. Such an approach would also acknowledge the core-neighborhood revitalization, population growth, and reinvestment that characterized much of the period between 1991 and 2006, as well as the growing diversity of developing suburbs and the greater economic and social differentiation taking place within suburbs. And it could do so through a variety of organizations and approaches within neighborhoods and cities and wherever the decisions shaping the processes of the globalizing economy take shape. Instead of pre-deciding the types of responses, community development approaches could explore the problem and trace the actors and processes that shape the problem. This might involve enhancing education within communities, negotiating for subsidized housing, and regulating national and global financial systems.

Conclusion: reclaiming neighborhoods from the inside out

In this article, we considered two scale mismatches: local–global and local–regional. Instead of becoming less relevant in a global world, neighborhoods have retained their importance as the places from which people experience, understand, and shape multi-scalar processes like globalization. We argued that the local and global are ever more intertwined and that a critical approach to community development means engaging the politics of place wherever those decisions take place. Instead of shifting attention to the region, refocusing energy on communities can help illuminate the processes that shape everyday lives and focus energy and attention on those processes. This means addressing immediate needs within communities as well as reaching beyond neighborhood boundaries to affect change. The regionalists have reached beyond the bounded localism of urban neighborhoods but they replaced the limited place of the neighborhood with the limited place of the region. Attempting to understand and resolve urban problems only within the place of metropolitan areas, in the context of a globalizing world that every day draws neighborhoods and globalizing processes closer together, distracts energy, organizations, and resources from exploring and understanding the many factors that shape urban change. Here we have focused on the financialization of the economy but that is but one element shaping urban change. There are many others but it is for community development literature and practitioners to explore the processes that shape conflict and develop their own response based on their own understanding of the problem from the perspective of their community. Community development has never been about just one approach nor has it involved one type of institution. What is needed now is an expansion of community development actors and approaches that uncover and engage with the politics of place wherever those politics are located.

In the 1960s and 1970s, people in urban neighborhoods across the United States realized that they did not have equal access to housing mortgage finance. They worked within their communities and then built a movement across the country achieving two major pieces of federal legislation—the Home Mortgage Disclosure Act (HMDA), which gathers information about home loan applications and originations, and the Community Reinvestment Act (CRA) which encourages depository financial institutions to do business in and with communities. Even as they secured this legislation, the world was changing. HMDA and CRA addressed the structures of the traditional banking system just as that system was coming undone. The globalization of finance and the financialization of the economy suggest new challenges. Community development actors can act within their communities and they can use their community experiences with finance and their desires to save and borrow in ways that benefit their communities to build local, national, and global organizations to craft new rules to use their savings in ways that benefit people every day. Such activities might involve revising global capital rules to reign in hyper-speculation and to strengthen consumer protection provisions (disclosure and financial literacy as the floor, not the ceiling). They might involve coordinating savings within communities to reinvest in communities gaining profit for savers and lower-risk, lower-cost investment for communities.

As community development moves into the future, tapping the historical roots of community development, which are about building access to democratic decision-making, and understanding the root causes of power imbalances is essential to under-stand the politics of place-making and to engage in it. To really engage with the processes that shape urban communities means looking beyond the boundaries of the neighborhood and the place of the region. It means better understanding the relation-ship between the neighborhood and the world. Community development needs to reassert its focus on building community-based democratic efforts to affect the politics that shapes everyday life.

Acknowledgments

Many thanks to Ronald van Kampen, Gideon Bolt, and Deborah Martin for their valuable comments and suggestions in shaping the article and to Robert Lake for offering a thoughtful critique.

Disclosure statement

No potential conflict of interest was reported by the authors.

References

Avery, Robert, Bhutta, Neil, Brevoort, Kenneth, & Canner, Glenn (2012). The mortgage market in 2011: Highlights from the data reported under the Home Mortgage Disclosure Act. *Federal Reserve Bulletin, 98*(6), 1–46.

Christophers, Brett (2014). *Geographies of finance II: Crisis, space and political-economic trans-formation. Progress in Human Geography.* Advance online publication. doi:10.1177/0309132513514343

Cisneros, Henry (1995). *Regionalism: The new geography of opportunity.* Washington, DC: U.S. Department of Housing and Urban Development.

Cox, Kevin R. (1998). Spaces of dependence, spaces of engagement and the politics of scale, or: Looking for local politics. *Political Geography, 17*(1), 1–23.

David, Imbroscio (2012). Beyond mobility: The limits of liberal urban policy. *Journal of Urban Affairs, 34*(1), 1–20.

DeFilippis, James (1999). Alternatives to the "New Urban Politics": Finding locality and auton-omy in local economic development. *Political Geography, 18*(8), 973–990.

DeFilippis, James (2004). *Unmaking Goliath: Community control in the face of global capital.* New York, NY: Routledge.

DeFilippis, James (2009). Alternatives to the "New Urban Politics": Finding locality and auton-omy in local economic development. *Political Geography, 18*(8), 973–990.

Demyank, Yulya, & Gopalan, Yadav (2007). *Subprime ARMs: Popular loans, poor performance.* Federal Reserve Bank of St. Louis. Spring. Retrieved from https://www.stlouisfed.org/publications/br/articles/?id=542

Dreier, Peter, Mollenkopf, John, & Swanstrom, Todd (2001). *Place matters: Metropolitics for the twenty-first century.* Lawrence, KS: University Press of Kansas.

Wyly, Elvin K., Atia, Mona, Foxcroft, Holly, Hammel, Daniel J., & Phillips-Watts, K. (2006). American home: Predatory mortgage capital and neighbourhood spaces of race and class exploitation in the United States. *Geografiska Annaler: Series B, Human Geography, 88*(1), 105–132.

Epp, Gayle (1996). Emerging strategies for revitalizing public housing communities. *Housing Policy Debate, 7*(3), 563–588.

Federal Reserve Bank of St. Louis. (2014). *30-year fixed rate mortgage average in the United States.* Retrieved from http://research.stlouisfed.org/fred2/series/MORTGAGE30US/#

Goetz, Edward G. (2003). *Clearing the way: Deconcentrating the poor in urban America.* Washington, DC: Urban institute Press. *AJS 112* (1), 231–275.

Goetz, Edward G., & Chapple, Karen (2010). 'You gotta move': Advancing the debate on the record of dispersal. *Housing Policy Debate, 20*(2), 209–236.

Gotham, Kevin Fox (2006). The secondary circuit of capital reconsidered: Globalization and the U. S. real estate sector. *American Journal of Sociology, 112*(1), 231–275.

Goodman, Laurie (2015). *The rebirth of securitization.* Urban Institute. Retrieved from http://www.urban.org/sites/default/files/alfresco/publication-pdfs/2000375-The-Rebirth-of-Securitization.pdf

Harvey, David (1987/1994). Flexible accumulation through urbanization: Reflections on 'Post-modernism' in the American City. In Ash Amin (Eds.), *Post-fordism: A reader* (pp. 361–386). Oxford: Blackwell.

Harvey, David (1989). *The limits to capital.* New York, NY: Verso.

Imbroscio, David (2008). "[U]nited and actuated by some common impulse of passion": Challenging the dispersal consensus in American housing policy research. *Journal of Urban Affairs, 30*(2), 111–130.

Imbroscio, David (2010). *Urban America reconsidered: Alternatives for governance and policy.* Ithaca, NY: Cornell University Press.

Jencks, Christopher, & Mayer, Susan E. (1990). The social consequences of growing up in a poor neighborhood. In Laurence E. Lynn Jr. & Michael McGeary (Eds.), *Inner city poverty in the United States* (pp. 111–186). Washington, DC: National Academy Press.

Lake, Robert (1994). Negotiating local autonomy. *Political Geography, 13*(5), 423–442.

Langley, Paul (2006). Securitising Suburbia: The transformation of Anglo-American mortgage finance. *Competition & Change, 10*(3), 283–299.

Lehmann, Nicholas (1994, January 9). The myth of community development. *New York Times.*

Leyshon, Andrew, & Thrift, Nigel. (2007). The capitalization of almost everything: The future of finance and capitalism. *Theory, Culture & Society, 24*(7–8), 97–115.

Macek, Steve (2006). *Urban nightmares: The media, the right, and the moral panic over the city.* Minneapolis: University of Minnesota Press.

Martin, Deborah (2004). Reconstructing urban politics: Neighborhood activism in land-use change. *Urban Affairs Review, 39*(5), 589–612.

Martin, Ron (2011). The local geographies of the financial crisis: From the housing bubble to economic recession and beyond. *Journal of Economic Geography, 11*(4), 587–618.

Massey, Douglas, & Denton, Nancy A. (1993). *American apartheid: Segregation and the making of the underclass.* Cambridge, MA: Harvard University Press.

McCoy, Patricia, Pavlov, Andrey, & Wachter, Susan (2009). Systemic risk through securitization: The result of deregulation and regulatory failure. *Connecticut Law Review, 41*(4), 493–591.

McGeary, Michael (1990). Ghetto poverty and federal policies and programs. In Laurence E. Lynn Jr. & Michael McGeary (Eds.), *Inner city poverty in the United States* (pp. 223–252). Washington, DC: National Academy Press.

Moreno, Louis (2014). The urban process under financialised capitalism. *City, 18*(3), 244–268.

Newman, Kathe, & Teresa, Benjamin (2013). *The emergent millennial urban politics of finance: Housing, regulation and informality in a post democratic world.* Paper presented at the 2013 Conference of the Urban Affairs Association.

Orfield, Myron (1997). *Metropolitics: A regional agenda for community and stability.* Washington, DC: Brookings Institution Press.

Pastor, Manuel, Brenner, Chris, & Matsuoka, Martha (2009). *This could be the start of something big: How social movements for regional equity are reshaping metropolitan America.* Ithaca, NY: Cornell University Press.

Pastor, Manuel, Peter Dreier, J, Grigsby, Eugene, III, & Marta, Lopez-Garza (2000). *Regions that work: How cities and suburbs can grow together.* Minneapolis: University of Minnesota Press.

Pierce, Joseph, Martin, Deborah, G., & Murphy, James T. (2011). Relational place-making: The networked politics of place. *Transactions of the Institute of British Geographers, 36*(1), 54–70.

Reeves, Jimmie, & Campbell, Richard (1994). *Cracked coverage: Television news, the anti-cocaine crusade, and the Reagan legacy.* Durham, NC: Duke University Press.

Rosenbloom, Robert A. (1981). The neighborhood movement: Where has it come from? Where is it going? *Nonprofit and Voluntary Sector Quarterly, 10*(4), 4–26.

Rusk, David (1993). *Cities without suburbs.* Washington, DC: The Woodrow Wilson Press.

Rusk, David (1999). *Inside game/outside game.* Washington, DC: Brookings.

Sadler, Bill, Wampler, Elizabeth, Wood, Jeff, Barry, Matt, & Wirfs-Brock, Jordan (2012). *The Denver regional equity atlas: Mapping access to opportunity at a regional scale.* Denver: Mile high connects. Retrieved from http://www.reconnectingamerica.org/resource-center/books-and-reports/2012/the-denver-regional-equity-atlas-mapping-opportunity-at-the-regional-scale/

Sassen, Saskia (2001). *The global city.* Princeton: Princeton University Press.

Sassen, Saskia (2005). The global city: Introducing a concept. *Brown Journal of World Affairs, 11*(2), 27–43.

Sharkey, Patrick (2013). *Stuck in place: Urban neighborhoods and the end of progress toward racial equality.* Chicago: University of Chicago Press.

Smith, Neil (1996). *The new urban frontier: Gentrification and the Revanchist city.* London: Routledge.

Swanstrom, Todd (1995). Philosopher in the city: The new regionalism debate. *Journal of Urban Affairs, 17*(3), 309–324.

Wilson, William J. (1987). *The truly disadvantaged: The inner city, the underclass, and public policy.* Chicago: University of Chicago Press.

The US Great Recession: exploring its association with Black neighborhood rise, decline and recovery

Derek Hyra[a] and Jacob S. Rugh[b]

[a]Department of Public Administration and Policy, School of Public Affairs, American University, Washington, DC, USA; [b]Department of Sociology, Brigham Young University, Provo, UT, USA

ABSTRACT

The United States experienced the Great Recession between 2007 and 2009 and many American cities and communities are still suffering from its legacy. During the prior period of the early and mid-2000s, many inner city African American communities were experiencing gentrification, driven in part by the real estate bubble that popped in 2007. While much has been written about the institutional and structural causes and consequences of the Great Recession, this article seeks to better understand its community-level implications by investigating the relationship between lending and property value patterns in three gentrifying African American communities just before, during and after this economic calamity. In particular, we investigate Bronzeville in Chicago, Harlem in New York City and Shaw/U Street in Washington, DC. Evidence suggests the Great Recession differentially influenced the development trajectories of these urban neighborhoods. In Bronzeville severe and prolonged property decline resulted, while much less economic stagnation was experienced in Harlem and Shaw/U Street. The Great Recession did not have uniform implications for urban African American neighborhoods: distinct community and city contexts, in particular racial and class neighborhood transitions and citywide unemployment and housing market conditions, mediate the influence of national economic decline and recovery.

Introduction

The US Great Recession (2007–2009) had a dramatic, and acute, short-term impact. In the years following the economic slowdown, property values across the country decreased nearly 30% (Gerardi, Foote, & Willen, 2011). Millions of homeowners found themselves in a state of owing mortgage lenders more than their homes were worth. Loan defaults and foreclosures mounted at unprecedented rates; between 2006 and 2010 approximately 14% of mortgages began the foreclosure process (Smith & Wachter, 2011). Many loans that defaulted had been bundled into secondary mortgage market financial products and the value of these securities plummeted (Keeley & Love, 2010; Levitin & Wachter, 2013). As a result credit markets tightened, economic activity nearly ceased, and unemployment skyrocketed from 5% to 10% in less than two years

(Hyra, Squires, Renner, & Kirk, 2013; Kirk & Hyra, 2012; Wachter & Smith, 2011). If not for nearly $14 trillion in federal government assistance (French, 2009), the US economy and its principal financial institutions might have completely collapsed.

Several encouraging signs indicate that the United States has, for the most part, survived one of the worst market downturns in its history. Since late 2009 the US economy has started to recover (Center on Budget and Policy Priorities, 2013; Wial, 2013). Foreclosure rates have significantly decreased, commercial and investment banks, as well as secondary mortgage market housing intermediaries, are netting billions in profits (Timiraos, 2013). In June 2014 national unemployment decreased to 6.1% (US Bureau of Labor Statistics, 2014) and property values have risen in many cities (Federal Housing Finance Agency, 2013; Whelan, 2013).

A plethora of studies have investigated the institutional and structural causes and consequences of this housing induced economic calamity. A number of investigations have explored the real estate bubble and the consequences of its burst. Some studies have explored the association between changing state and national financial regulatory policies and the rise in subprime lending (Avery & Brevoort, 2011; Bostic, Engel, McCoy, Pennington-Cross, & Wachter, 2008; Engel & McCoy, 2011; Gotham, 2009; Gramlich, 2007; Immergluck, 2009; Levitin & Wachter, 2013). Others have investigated the relationship between metropolitan segregation and its association with patterns of subprime lending and foreclosures (Anacker, Car, & Pradhan, 2012; Been, Ellen, & Madar, 2009; Hyra et al., 2013; Rugh & Massey, 2010). Another set of studies and reports have focused on the various financial impacts of the collapse on city (US Conference of Mayors, 2007) and state (Joint Economic Committee, 2007) budgets, as well as on the household wealth of different racial and ethnic groups (Shapiro, Meschede, & Osoro, 2013; Taylor, Kochhar, Fry, Velasco, & Motel, 2011). Lastly, scholars have assessed the massive federal government response to stabilize key financial institutions as well as minority communities suffering from foreclosure concentration (French, 2009; Immergluck, 2013; Squires & Hyra, 2010).

While there have been many investigations of the boom and bust periods associated with the Great Recession, few studies have investigated the post-recession recovery period (with the exception of Wial, 2013) and its influence on urban African American neighborhoods. It is important to study Black communities because these areas, compared to White neighborhoods, had a disproportionate number of subprime loans (Avery, Brevoort, & Canner, 2007; Been et al., 2009; Bunce, Gruenstein, Herbert, & Scheessele, 2001; Calem, Hershaff, & Wachter, 2004) and suffered unduly from foreclosure concentration once the recession hit (Rugh & Massey, 2010; Rugh, Albright, & Massey, 2015).

Despite recent signs of national economic recovery, many urban Black neighborhoods are still feeling the lingering negative effects of the Great Recession. For instance, some neighborhoods continue to be dotted with foreclosed properties and have not experienced property value upticks (Dreier, Bhatti, Call, Schwartz, & Squires, 2014). However, other communities have had an increase in commercial and residential investments and rising property values, making the Great Recession seem like a distant past.

It is important to understand how Great Recession dynamics, particularly subprime lending and foreclosures, affected urban African American neighborhoods over time (Li

& Morrow-Jones, 2010). We know little about this topic because many case studies of inner city revitalization in the 2000s failed to sufficiently account for high cost lending (e.g., Boyd, 2008; Freeman, 2006; Hyra, 2008; Pattillo, 2007). While some more recent studies have demonstrated the subprime and foreclosure devastation that occurred in African American, and other minority, neighborhoods just before and during the recession (Hyra et al., 2013; Immergluck, 2010; Rugh & Massey, 2010), few, if any, have explored what happened in these communities as the national economy began to recover. This study contributes to and extends the Great Recession literature by investigating the ways in which three Great Recession-related periods: the boom (2000–2006), bust (2007–2009) and recovery (2010–2012) relate to the development trajectories of Black inner city neighborhoods.

By studying the development trajectories of three historically African American neighborhoods, Bronzeville in Chicago, Harlem in New York City, and Shaw/U Street in Washington, DC, this study provides a better understanding of the relationship between the Great Recession and inner city neighborhood change. We assess the developmental trajectories of these neighborhoods over a 12-year period (2000–2012) in three distinct Great Recession phases. While all three communities followed the same general pattern of a boom and bust, our findings suggest that the race and class of new homeowners in these communities and citywide unemployment and housing market trends were critical to understanding distinct community-level, post-recession recovery. This exploratory study highlights that the Great Recession did not have uniform implications for urban African American neighborhoods and suggests that city and community contexts mediate the impact of national economic decline and recovery.

Neighborhood change and the Great Recession

Macro- and micro-level forces originating beyond and within neighborhoods are associated with their changing conditions. Macro dynamics, such as the global economy (Sassen, 2012; Sites, 2003), the national economy, and federal policy directives (Halpern, 1995; Peterson, 1981), are important factors that shape urban neighborhoods. Furthermore, city-level economic and political factors, such as housing market conditions and political actions, mediate global and national forces to influence neighborhood change (Aalbers, 2011; Hirsch, 1998; Hyra, 2008; Logan & Molotch, 2007; Stone, 1989). Not only are multiple external neighborhood dynamics important for understanding community change, but internal neighborhood circumstances, such as organizational structure (Wilson, 1996) and collective efficacy, that is neighborhood norms of trust and collective action (Sampson, Raudenbush, & Earls, 1997), influence the neighborhood change process. For instance, collective efficacy is associated with levels of neighborhood crime, which is often associated with community-level investment and population movement patterns that in turn can influence property values (Hwang & Sampson, 2014; Sampson, 2012; Taub, Taylor, & Dunham, 1987).

Some studies have investigated how different Great Recession-related dynamics influenced neighborhood change. Prior to the financial collapse, scholars argued that in the 2000s the United States experienced a new round of urban renewal (Hyra, 2012), as many inner city Black communities that had been entrenched in concentrated poverty for almost 50 years revitalized and transitioned to more mixed-income environments (Boyd, 2008; Coleman, 2012; Freeman, 2006; Hyra, 2008; Pattillo, 2007; Ruble, 2010). While several processes and national policies, such as the Housing Opportunities for People Everywhere (HOPE VI) program (Goetz, 2013; Vale, 2013), relate to this circumstance, some scholars claim that increases in subprime, or high cost, lending were a major factor (Hyra, 2012; Maeckelbergh, 2012; Wyly, Moos, Hammel, & Kabahizi, 2009).[1]

After decades of mortgage loan denials and the redlining of African American communities (Massey & Denton, 1993), in the 2000s mortgage capital finally found "an inner city fix" (Wyly, Atia, & Hammel, 2004). Some studies claim that national policies changes in the financial regulatory framework, such as risk-based pricing, relate to the proliferation of subprime loans in inner city areas and ensuing community revitalization (Hyra, 2012; Wyly et al., 2004). These investigations suggest that the run up in subprime lending, along with other factors, helped to gentrify, and prop up property values in, certain inner city Black neighborhoods.

The greenlining of credit in minority communities was seen as controversial (Aalbers, 2011; Squires, 2005, 2008).[2] On the one hand, it provided minorities, who had once been denied loans, the ability to buy a home and attain the American Dream.[3] Furthermore, it was associated with the revitalization of historic, once divested, Black communities (Hyra, 2012). However, the dream quickly turned into a nightmare for many African Americans as a disproportionate percentage of loans originated in African American communities had high interest rates or other subprime features, such as introductory teaser rates, prepayment penalties and balloon payments (Avery, Brevoort, & Canner, 2006; Been et al., 2009; Wyly et al., 2009). Several studies demonstrate that subprime loans, compared to prime loans, disproportionate default (Quercia, Stegman, & Davis, 2007) and lead to foreclosed properties (Immergluck & Smith, 2005). Many of these high-priced, subprime loans originated in African American communities were unsustainable and ultimately ended in default, stripping borrower equity and lining many Black communities with foreclosed properties. Homeownership has declined among African Americans more than any other racial or ethnic group (Kuebler & Rugh, 2013). Some estimate that the Great Recession wiped away half of Black America's wealth (Shapiro et al., 2013) and contributed to widening the wealth gap between Blacks and Whites (Taylor et al., 2011). Some claim that subprime lending in African American communities was a devastating form of "reverse redlining" (Squires, 2008; Williams, Nesiba, & McConnell, 2005).

Black neighborhoods that have high proportions of subprime loans likely have high foreclosure concentrations. Foreclosures can be extremely problematic for several reasons. First, abandoned properties can result in significant residential turnover (Li & Morrow-Jones, 2010; Kirk & Hyra, 2012) and property value decline in adjacent homes (Anenberg & Kung, 2012; Gerardi, Rosenblatt, Willen, & Yao, 2012; Immergluck & Smith, 2006a; Lin, Rosenblatt, & Yao, 2009). Furthermore, some studies suggest that foreclosures are associated with other social costs, such as increased levels of neighborhood crime, which might also encourage more people to leave a neighborhood and further drag down property values (Ellen, Lacoe, & Sharygin, 2013; Immergluck & Smith, 2006b).[4]

While subprime loans and foreclosures are disproportionately in African American neighborhoods, their levels are likely unequally distributed among all inner city urban Black communities. First, cities with elevated Black/White segregation have, on average, a higher proportion of subprime loan originations (Hyra et al., 2013) and foreclosures (Rugh & Massey, 2010). Second, cities with stronger economies and lower unemployment rates might have lower rates of subprime lending. Third, the types of newcomers that select into gentrifying Black neighborhoods might influence their propensities to have subprime and foreclosure concentrations. For instance, African American neighborhoods experiencing an influx of Black middle class (income $50,000–$100,000), as opposed to elite Black ($200,000+), residents might be more vulnerable to subprime loan and foreclosure concentrations since those with ample income might qualify for prime loans. Additionally, if middle- and upper-income Whites move to gentrifying African American neighborhoods, we would expect these communities to have fewer subprime loans and foreclosures, compared to more homogeneous Black communities. Community differences in subprime and foreclosure levels, both during the pre-recession and recession years, might relate to property value resiliency during the post-recession recovery years.

Methodology and design

This longitudinal multiple case study design (Yin, 2013) explores a variety of quantitative datasets to investigate the development trajectories, between 2000 and 2012, of three inner city African American neighborhoods located in three different cities. We segmented this 12-year period into three distinct phases: pre-Recession (2000–2006), Recession (2007–2009) and post-Recession (2010–2012) periods. In each phase we assessed Home Mortgage Disclosure Act (HMDA) neighborhood-level lending data, such as the number of home loans and dollar amounts, the percent and dollar amount of high cost loans, median borrower income, and borrower race. We also assessed community-level home values as well as foreclosure rates. Lastly, we evaluated decennial census demographic information, such as the population level and racial composition, of each neighborhood, and citywide circumstances including unemployment and housing market conditions.[5] These data were used to investigate how certain African American neighborhoods faired over time prior to, during and after the Great Recession.

We chose to compare Bronzeville in Chicago, Harlem in New York City, Shaw/U Street in Washington, DC.[6] These neighborhoods were selected since they were, and to some extent still are, the African American hubs of their respective cities. Furthermore, they all experienced gentrification during the 2000 boom years (see Boyd, 2008; Freeman, 2006; Hyra, 2008; Ruble, 2010). They also had relatively similar development histories, which minimizes, but does not eliminate, the chance that distinctions in their 12-year Great Recession-related trajectories result from undetected historical differences among these neighborhoods.

Bronzeville, Harlem and Shaw/U Street share many characteristics as iconic African American neighborhoods. Much of each neighborhood's older housing stock was constructed in the mid- to late-nineteenth century when these areas were mainly affluent and middle-class White enclaves (Mahoney, 2001; Osofsky, 1996; Spear, 1967; Williams, 2002). Following the Great Migrations of African Americans from the South (Grossman, 1989), these neighborhoods became predominantly mixed-income African American districts during the 1920s, 1930s and 1940s (Drake & Cayton, 1945; Osofsky,

1996; Ruble, 2010). These racial enclaves became "cities within cities," as African Americans were prevented from moving to other parts of these cities, in part, due to restrictive covenants and racial violence (Hirsch, 1998; Massey & Denton, 1993). The end of restrictive covenants in the late 1940s was associated with an African American middle-class exodus from these communities during the 1950s, 1960s and 1970s. During the second half of the twentieth century, these areas became concentrated with subsidized housing (Hirsch, 1998; Hyra, 2008; Ruble, 2010). The construction of subsidized housing, combined with national deindustrialization and African American job loss, resulted in these communities becoming extremely impoverished areas (Clark, 1965; Liebow, 1967; Wilson, 1996). In the 1980s and 1990s, each of these neighborhoods were "no go" zones with extremely high levels of poverty and crime (Robinson, 2010; Taylor, 2002; Venkatesh, 2000). In the 1990s these communities started to revitalize with increased public and private investments and in the 2000s their property values began to skyrocket during the subprime lending boom.

While each of these neighborhoods was predominately African American in late 1990s and early 2000s, each experienced a slightly different type of gentrification. In the 1990s Bronzeville and Harlem experienced mainly Black gentrification (Hyra, 2008), while Shaw/U Street had a much greater number of Whites as well as upper income African Americans move to the neighborhood (Ruble, 2010). In 1990, Bronzeville, Harlem and Shaw/U Street were 95%, 88% and 67% Black, respectively. By 2000, their Black population percent had declined to 92% in Bronzeville, 77% in Harlem and 52% in Shaw/U Street (see Maps 1–3 and Table A1). Changing neighborhood racial composition is important since this might be associated with different subprime lending

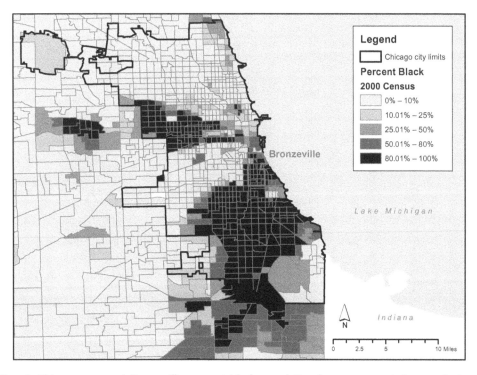

Map 1. Chicago area and Bronzeville percent black population by census tract. Source: Authors' calculations based on 2000 US Census data.

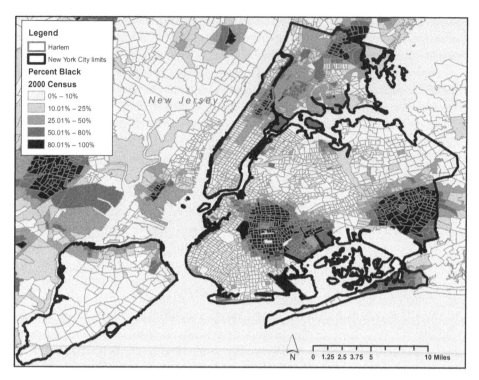

Map 2. New York City area and Harlem percent black population by census tract. Source: Authors' calculations based on 2000 US Census data.

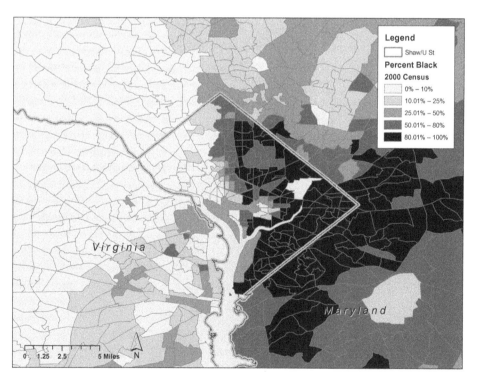

Map 3. Washington, DC area and Shaw/U St. percent black population by census tract. Source: Authors' calculations based on 2000 US Census data.

rates between 2000 and 2006, foreclosure rates during the recession, and real estate stabilization during the recovery.

These communities were embedded within cities with different economies. While all were considered global cities, New York is one of the world's financial sector power-houses (Sassen, 2012; Sites, 2003), Washington, DC, as the nation's capital, is dominated by the white-collar federal government employment (Abbott, 1999; Gillette, 1995), and Chicago has a mixed economy consisting of white-collar and blue-collar workers (Bennett, 2010; Ranney, 2003). These different city-level economies might have been differentially impacted by the Great Recession, which could have implications for the level of decline and type of post-recession recovery in their respective African American neighborhoods.

Findings

In the remainder of the article, we assess the relationship between the Great Recession-associated dynamics, in the pre-recession, recession, and post-recession periods, and the development trajectories of three iconic African American communities. As Figure 1 shows, the property values in Bronzeville, Harlem and Shaw/U Street follow a similar pattern during the first two periods, a run up during the boom years (2000–2006) and a decline during the recession (2007–2009). However, near the end of the Great Recession recovery period (2010–2012), Harlem and Shaw/U Street's property values exceed their boom period heights, while Bronzeville's home values continue to depreciate. We employ a variety of data to help explain, and generate hypotheses, related to these divergent, post-recession property value trajectories.

The boom years, 2000–2006: subprime lending and gentrification

As noted, during the 2000s several low-income, inner city African American neighbor-hood redeveloped around the country (Hyra, 2012). This period saw a national rise in subprime lending, much of which was disproportionately received by African Americans living in predominately Black communities. Between 2000 and 2006, the percent of total mortgage originations that were subprime tripled, going from 12% to 36% (Engel & McCoy, 2011). In 2006, at the height of subprime lending, 54% of African American, 47% of Hispanic and 18% of White mortgage borrowers received a high cost loan (Avery et al., 2007). Moreover, in census tracts where the population was at least 80% minority, 47% of borrowers obtained high-priced loans, compared with 22% of borrowers in communities where racial and ethnic minorities accounted for less than 10% of the population.

During the subprime boom years, between 2000 and 2006, Bronzeville, Harlem and Shaw/U Street experienced the greenlining lending effect. In 2000, in Bronzeville, there were just 814 home loans made but by 2006 the number had jumped to 2,049, a 152% increase. In Harlem, 394 mortgages loans were made in 2000 and in 2006 the number increased to 733, an 86% jump. In Shaw/U Street the story was much the same, between 2000 and 2006, the community experienced a 133% increase in mortgage lending (see Table A2).

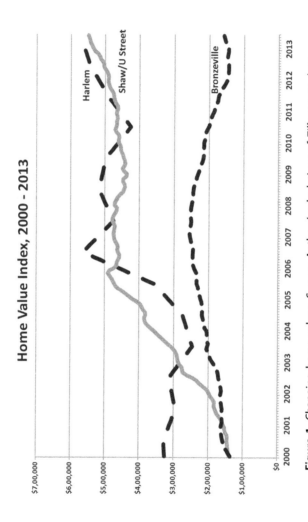

Figure 1. Changing home values. Source: Authors' calculations of Zillow.com data.

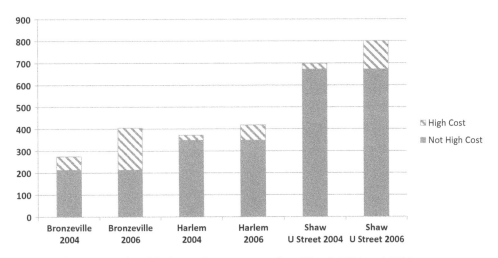

Figure 2. Changing total and high cost loan amounts (in millions), 2004 and 2006.

The increased home lending in these African American communities was associated with rising subprime originations.[7] Figure 2 shows the simultaneous increase in total (*T*) and high cost (HC) loan dollar amounts in 2004 and 2006 in each community. In these communities total home lending from 2004 to 2006 increased from $1.34 billion to $1.86 billion and high cost lending rose from $111 million to $390 million (see Table A3).

While total and high cost lending increased dramatically in these communities during 2004 and 2006, there were distinct subprime saturation levels. In Bronzeville, the percent of high cost home loan originations in Bronzeville doubled, going from 23% to 48%, a 109% increase. In Harlem, subprime lending increased from 6% to 12%, a 100% increase. In Shaw/U Street, during the same period, high cost lending increased from 4% to 15%, a 275% increase (see Table A4).

The larger context of neighborhood high cost lending for the regions surrounding Bronzeville, Harlem and Shaw/U Street are displayed in Maps 4–6. These maps show the high cost proportion of home loans made from 2004 to 2007. They further under-score the differential saturation of high cost loans across the three case study commu-nities. In Shaw/U Street and Harlem, neighborhood rates of high cost lending were substantially lower relative to other similarly predominantly Black neighborhoods in their respective regions. In Bronzeville, however, the rates of high cost lending were far greater and in line with other mostly Black neighborhoods in the greater Chicago region.

The boom in high cost lending was associated with property value escalation in all three communities between 2000 and 2006. In Bronzeville, property values increased 75%, from $150,000 to $263,000. Harlem's property values increased 61%, from $324,000 to $522,000. And in Shaw/U Street, single-family home price increased from $189,104 to $629,950, a staggering 233% increase.

While all three communities experienced increased prime and subprime lending and property values during the pre-recession boom period, they experienced different types of gentrification. Bronzeville experienced Black gentrification. From 2004 to

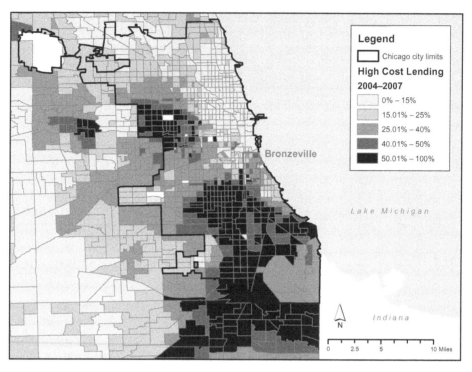

Map 4. Chicago area and Bronzeville high cost lending rates by census tract. Source: Authors' calculations based on HUD Neighborhood Stabilization Program Round 3 data.

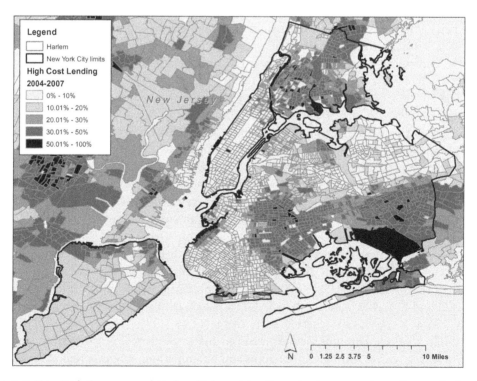

Map 5. New York City area and Harlem high cost lending rates by census tract. Source: Authors' calculations based on HUD Neighborhood Stabilization Program Round 3 data.

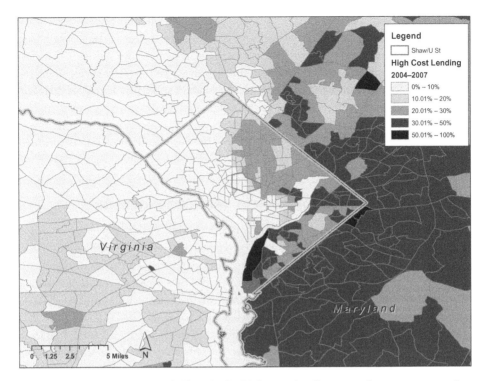

Map 6. Washington, DC area and Shaw/U St. high cost lending rates by census tract. Source: Authors' calculations based on HUD Neighborhood Stabilization Program Round 3 data.

2006, the median home borrower income increased from $73,000 to $86,000 and the percent of African American homebuyers increased from 70% to 74%, signaling that the community was mainly attracting new Black middle-class members (see Table A5).

In Harlem and Shaw/U Street, during the pre-recession period, multiracial gentrification occurred with an influx of upper income African American and White homeowners. Between 2004 and 2006, the percent of Harlem homebuyers making over $200,000 a year increased from 11% to 31%. While several hundred of these new elite Harlem homebuyers were African American, many were also White. Between 2004 and 2006, the percentage of White Harlem homebuyers increased, while mortgage originations to African American declined from 55% to 51%, suggesting that Harlem was experiencing a racial transition, at least among its new home buying population (see Table 2).[8]

Shaw/U Street, like Harlem, experienced a racial and class transition, suggesting that the community was also experiencing multiracial gentrification. Between 2004 and 2006, the percent of mortgage borrowers with incomes over $200,000 increased from 12% to 19% (see Table A6). While the percent of Shaw/U Street Black borrowers increased slightly between 2004 and 2006, in 2006 24% of those obtaining mortgage loans in the community were African American and 60% were White, indicating a mixed-race homebuyer influx during the pre-recession boom years.

There were important similarities and differences among Bronzeville, Harlem and Shaw/U Street during the pre-recession years. All three communities experienced sharp rises in overall mortgage lending, both prime and subprime, and their property values skyrocketed. However, homebuyers in these communities differed. In Bronzeville, the new homeowners tended to be middle-income African Americans, while in Harlem and Shaw/U Street new mortgage holders were more affluent and racially mixed. There were also major differences among these communities in the subprime loans. In 2006, at the height of subprime lending, 48% of loans originated in Bronzeville were high cost, compared to just 12% in Harlem and 15% in Shaw/U Street. These differences in subprime lending rates might help explicate how the real estate bubble crash influenced these communities.

The crash: disinvestment, foreclosures and unemployment

In December 2007, the US national economic slowdown officially began. By this time home values were on the decline, brought on partly by mounting defaults on unsustainable high cost home loans. Once home values declined, subprime borrowers were deemed too risky by banks to refinance toward lower interest rates because their properties had high loan-to-value ratios. Many of these unlucky subprime borrowers defaulted, dotting minority communities with foreclosures. The housing market bust was compounded by the fact that several major Wall Street investment houses held rapidly devaluing mortgage-backed securities. In September 2008, Lehman Brothers, a major Wall Street investment bank, went bankrupt after it could not cover loan-related losses. This triggered a credit freeze, which made it very difficult for individuals and companies to get new loans or refinance existing ones. This credit freeze drastically slowed the overall economy, and resulted in a rising unemployment rate. Increased unemployment led to further loans defaults, foreclosures and deep dips in home prices.

The Great Recession immediately affected Bronzeville, Harlem and Shaw/U Street; however, there were variations in overall mortgage lending and subprime activity, foreclosures and property declines. In Bronzeville, 2,049 home loans were made in 2006 and by 2008 that number had decreased to 658, a 68% decline. In Harlem and Shaw/U Street the total mortgage lending decreased by 8% and 41%, respectively.

During the recession there were huge drop-offs in high cost, subprime lending. Figure 3 displays the changing dollar amounts (in millions) of high cost loans in each community between 2006 and 2010. Between 2006 and 2008, Bronzeville's subprime lending amount decreased from $192 million to $29 million, Harlem's from $69 million to $12 million and Shaw/U Street's from $129 million to $11 million. Clearly, by 2008, the subprime greenlining to these African American communities had ended.

Mortgage lending decreased, foreclosure rates climbed, and property values fell. Between 2004 and 2008, the foreclosures rates increased 300%, 500% and 425% in Bronzeville, Harlem and Shaw/U Street, respectively. As seen in Table 1, by 2008, Bronzeville's foreclosure rate was 8.1%, Harlem's was 1.8%, and Shaw/U Street was 2.1%.

We place the foreclosure rates in the case study communities in the larger regional contexts of their respective metropolitan regions in Maps 7–9. The maps show the percent of all home loans made from 2004 to 2007 that started the foreclosure process during 2009 and 2010 as calculated using data from the US Department of Housing and Urban Development's Round 3 implementation of the Neighborhood

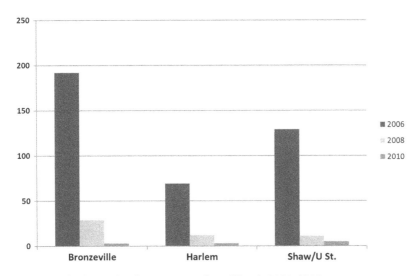

Figure 3. Changing high cost lending amounts (in millions), 2006–2010.

Table 1. Foreclosure rate (values in %).

	2004	2006	2008	2010	2012
Bronzeville	2.0	6.3	8.1	9.0	6.1
Harlem	0.3	1.5	1.8	2.6	0.7
Shaw/U Street	0.4	0.8	2.1	2.0	0.4

Source: HUD NSP1 (2008), HUD NSP3 (2010), Furman Center for Real Estate & Urban Policy, Urban Institute, Woodstock Institute, CoreLogic, and authors' calculations. Note: Foreclosure starts/filing rate. Denominator is mortgaged properties.

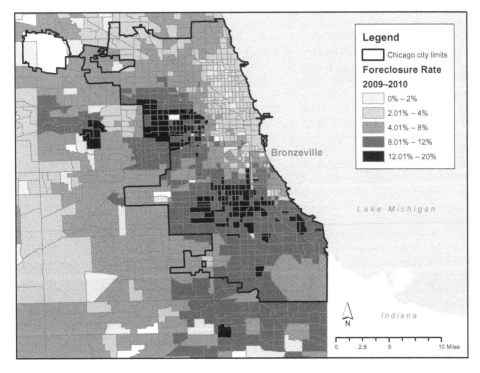

Map 7. Chicago area and Bronzeville foreclosure start rates by census tract. Source: Authors' calculations based on HUD Neighborhood Stabilization Program Round 3 data.

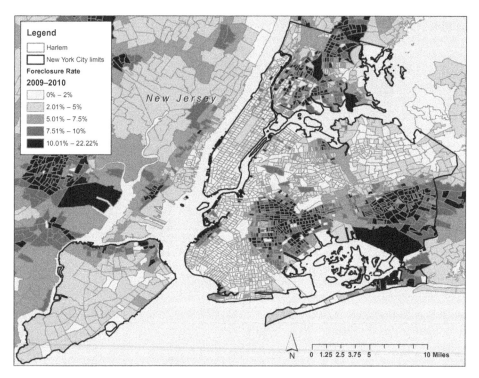

Map 8. New York City area and Harlem foreclosure start rates by census tract. Source: Authors' calculations based on HUD Neighborhood Stabilization Program Round 3 data.

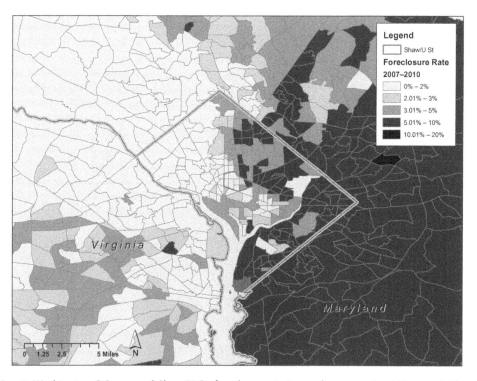

Map 9. Washington, DC area and Shaw/U St. foreclosure start rates by census tract. Source: Authors' calculations based on HUD Neighborhood Stabilization Program Round 3 data.

Stabilization Program (NSP). During the worst of the foreclosure crisis, nearly all the census tracts in Bronzeville experienced foreclosure start rates of 8–12%. In contrast, nearly all tracts in Harlem and Shaw/U Street exhibited foreclosure start rates under 5% with only one exception in each of those areas. Overall these maps further emphasize the extent to which Harlem and Shaw/U Street escaped the worst of the foreclosure crisis while Bronzeville was not spared from the elevated level of foreclosures across most predominantly Black neighborhoods in all three metropolitan areas.

Property values declined at differential rates in these neighborhoods. In Bronzeville, between 2007 and 2009, home values fell from $263,350 to $210,200, a 20% decline. In contrast, Harlem's property values remained almost completely stable over the 2-year recession period, declining from $492,677 to $492,529. Shaw/U Street's recession period property value decline, from $472,588 to $447,213, or 5%, was not as severe as Bronzeville's nor was it as moderate as the circumstances in Harlem.

The types of gentrification in these neighborhoods during the pre-recession boom period and recession years help explain distinct community-level foreclosure rates and property value declines. As noted, during the pre-recession boom years, Harlem and Shaw/U Street had lower subprime lending rates and experienced an influx of more affluent and racially mixed homeowners, compared to Bronzeville. As indicated in Table 2, the trend of increased White investment in Harlem and Shaw/U Street continued during the recession years. Between 2008 and 2010, Harlem's percent of White borrowers rose from 40% to 46%, while the percent of Black borrowers decreased from 32% to 28%. A similar home investment trend occurred in Shaw/U Street; White home borrowing rose from 75% to 79%, while Black home borrowing decreased from 13% to 9%. In Bronzeville, the percent of home loans received by Whites decreased from 26% to 24% and African Americans continued to obtain the majority of home loans during the Great Recession.

Community-level foreclosure rate and property value differences might have also been due to variations in citywide unemployment and housing market conditions.[9] Table 3 shows the changing citywide unemployment rates for Chicago, New York and Washington, DC between 2004 and 2012. Between 2006 and 2008, Chicago had the

Table 2. Home loan borrower race/ethnicity percent by community and year (values in %).

	2002	2004	2006	2008	2010	2012
Bronzeville						
Black	73	70	74	67	66	77
White	19	21	18	26	24	11
Latino	3	4	5	3	4	7
Asian/PI	6	5	2	5	7	5
Harlem						
Black	55	55	51	32	28	23
White	33	33	40	46	53	60
Latino	7	7	6	9	7	7
Asian/PI	5	5	4	13	13	11
Shaw/U St.						
Black	–	22	24	13	9	8
White	–	66	60	75	79	79
Latino	–	5	7	4	4	5
Asian/PI	–	7	8	8	8	9

Source: HMDA Loan Application Register (excludes missing race).

Table 3. Citywide unemployment rates (values in %).

	2004	2006	2008	2010	2012
Chicago	7.5	5.3	6.9	11.6	10.1
New York	7.1	5.0	5.5	9.6	9.2
DC	8.2	6.0	7.0	9.9	8.9

Source: US Bureau of Labor Statistics.

Table 4. Metropolitan homeownership vacancy rates (values in %).

	2006	2009	2012
Chicago	2.1	3.8	2.8
New York	1.7	2.7	2.1
DC	1.6	2.2	1.7

Source: US Census 2014.

largest increase in unemployment, moving up 1.6%. During the same period, in New York and Washington, DC unemployment increased by 0.5% and 1%, respectively. Thus, from an employment standpoint the Great Recession hit Chicago the hardest and this might also help to explain why Bronzeville's 2008 foreclosure rate was elevated compared to Harlem and Shaw/U Street.

The cities investigated also had slightly different housing market conditions (see Table 4). Chicago's housing market, compared to New York and Washington, DC, was looser and had more slack (i.e., a higher vacancy rate). In 2006, the Chicago metropolitan statistical area (MSA) had a homeownership vacancy rate of 2.1% compared to 1.7% and 1.6% in New York and Washington, DC metros, respectively (US Census Bureau, 2014). Moreover, during the recession the vacancy rate increased the most in Chicago (1.7%) compared to New York (1%) and Washington, DC (0.6%). The vacancy data suggest that there was more initial slack in Chicago's housing market, which might have contributed to Bronzeville's type of pre-recession gentrification, its recession period property value decline, and the community's lackluster post-recession recovery.

The Recovery

In 2010 the United States began to recover from the recession as economic activity and property values increased and foreclosures and unemployment decreased nationwide. However, this recovery was not equally felt across our case study communities. Even in 2012, Bronzeville had few signs of an economic recovery; the foreclosure rate remained relatively high and property values continued to decline. Harlem and Shaw/U Street foreclosures were, in contrast, almost nonexistent and property values had surpassed their 2006 heights.

While Harlem and Shaw/U Street rebounded from the Great Recession, Bronzeville remained dotted with foreclosures during the recovery period. The foreclosure rate actually increased from 2008 to 2010, going from 8.1% to 9% before eventually moving down to 6.1% in 2012. By comparison, in 2012, Harlem and Shaw/U Street's foreclosure rates were under 1%. As the nation began to recover between 2008 and 2010, Bronzeville's median borrower income level dropped from $84,000 to $79,000. Furthermore, the community substantially lost a quarter of its population between 2000 and 2010 and remained racially homogenous (see Table A7).

While Bronzeville continued to suffer from the legacy of the Great Recession, between 2010 and 2012, Harlem and Shaw/U Street were experiencing gentrification overdrive. In these communities, the pre-recession and recession trends of a more racial diverse set of upper-income homebuyers continued during the recovery years. Between 2010 and 2012, Harlem's White home loan borrowers increased from 53% to 60% and African Americans decreased from 28% to 23%. In Shaw/U Street, the number of home loans in 2008 was 1,334 and in 2012 it increased to 2,312, of which 79% were originated to Whites and only 9% were originated to African Americans.

By 2012, Harlem and Shaw/U Street could no longer be characterized as Black ghettos but were better described as trendy mixed-race, gilded ghettos. In the recovery period, Harlem and Shaw/U Street became lined with trendy restaurants, beer gardens and wine bars, and were arguably among the hippest neighborhoods in their respective cities. Some of New York City's most sought after restaurants, such the Red Rooster, opened in Harlem and Shaw/U Street's 14th Street corridor, one of the community's main thoroughfares, boasted over 24 new upscale, chic restaurants, many of which opened between 2010 and 2012 (see Collins, 2010; Shin, 2013; Wessel, 2013). In Harlem and Shaw/U Street, the Great Recession was unrecognizable, while its harsh legacy continued to loom over Bronzeville.

Discussion

By exploring lending patterns in three historic African American neighborhoods during three distinct Great Recession-related periods, this study demonstrates several important findings. First, the real estate bubble and subprime lending in the early- and mid-2000s were associated with the gentrification of Black inner city neighborhoods. During the 2000–2006 bubble period a significant amount of lending, much of which was subprime, increased over time in Bronzeville, Harlem and Shaw/U Street. In these neighborhoods, increased high cost lending coincided with skyrocketing property values. While this study does not rigorously isolate the effect of high cost lending on property values, when the recession hit and subprime lending percentages were greatly reduced, Bronzeville, Harlem and Shaw/U Street's property values plummeted, suggesting that the pre-recession rise in subprime lending, in part, helped to temporarily revitalize these African American inner city neighborhoods.

While highlighting the link between subprime lending and the revitalization of urban African American neighborhoods, this comparative study also suggests that certain types of gentrification, along race and class lines, shielded some redeveloping African American neighborhoods from the full external Great Recession shock as well as facilitated post-recession recovery. Harlem and Shaw/U Street's elite upper class, mixed-race gentrification, compared to Bronzeville's Black middle-class gentrification, might have protected these communities from excessive subprime lending rates and foreclosure concentrations, which could explain why these areas recovered more quickly during the post-recession period. After 2012, Harlem and Shaw/U Street's Great Recession property values were completely restored, while Bronzeville's home values continued to decline.

The divergent post-recession trajectories of Bronzeville, Harlem and Shaw/U Street suggest that middle-class African Americans, and certain communities they move to, remain vulnerable to fiscal shocks, compared to the communities where more elite African Americans reside. This finding supports some of the research of Karyn Lacy (2007, 2012) and Mary Pattillo (2007) in that it demonstrates that there are important class differences within Black America and the Great Recession's influence on African Americans and the communities in which they reside needs a class-based analysis to be fully understood. For instance, the Black middle class that moved to Bronzeville is still suffering from property value loss, while the elite African Americans who relocated to Harlem and Shaw/U Street have experienced, for the most part, post-recession home value recovery. There are important and meaningful distinctions within Black America and certain Black middle-class individuals and communities remain more financially vulnerable to fiscal shocks than others.

While class differences within Black America are important to understanding the long-term implications of the Great Recession, this study also suggests that White settlement patterns remain important for understanding community change processes. The influx of Whites into certain urban African American communities occurred prior to, during, and after the Great Recession. The movement of Whites to African American communities was associated with post-recession property value increases. The one community that stayed primarily African American during the downturn continued to have depressed property values during the recovery period. Knowing the vast wealth disparities that remain between Whites and Blacks (Shapiro, 2004), it is not surprising that a White influx into Black gentrifying communities was associated with lower subprime and foreclosure rates and quicker community-level property value revival. Thus, this research suggests that race and class analyses are important to understanding the longer term implications of the Great Recession in African American communities.

Beyond race and class demographic transitions, it seems metropolitan context is another important variable for understanding African American community change processes associated with the Great Recession. The elevated subprime and foreclosure rates in Bronzeville could be related to the broader Chicago economic context compared to New York City and Washington, DC. Citywide unemployment and home-ownership vacancy data suggest that the Great Recession hit Chicago relatively hard compared to New York and Washington, DC and this citywide impact might help to explain the continued downward trajectory of Bronzeville, compared to Harlem and Shaw/U Street during the recovery period.

This study is one of the first to investigate the pre-Great Recession, Great Recession and post-Great Recession periods to understand the ways in which one of America's most drastic economic calamities is associated with inner city African American neighborhood change. The study's main aim was to better comprehend divergent post-recession neighborhood property value patterns in African American neighborhoods experiencing gentrification during the pre-recession real estate boom. While this study has several interesting findings, it is limited and should be evaluated based on its ability to generate new and meaningful Great Recession-related community change hypotheses to be further tested in subsequent studies with more rigorous quantitative methods.

This study suggests that future neighborhood change research needs to better understand how the types of gentrification experienced before and during the Great Recession mediate its overall community-level impact. While this study attempted to explore how White and Black community influx mediated Great Recession impacts through community-level subprime rates and foreclosure concentration, future studies need to better define and quantify different types of gentrification. A related limitation is that this study almost exclusively focused on new homeowner influx as one of the prominent mediating variables between Great Recession dynamics and property values. To more fully unpack determinants of property value decline and recovery, future Great Recession studies should assess changing renter profiles, as many urban African American neighborhoods have more renters than homeowners. Another important neighborhood change variable is internal community circumstances, such as levels of collective efficacy or organizational-led revitalization efforts (Sampson, 2012; Taub et al., 1987). This study made no attempt to document variations in the neighborhood rental markets or community-driven improvement initiatives and these factors might contribute to divergent redevelopment trajectories.

While this study highlighted citywide unemployment and metropolitan housing market conditions as other mediating variables, additionally citywide dynamics, related to subprime lending and foreclosure rates, need to be assessed. Studies have demonstrated that metropolitan segregation relates to patterns of high cost lending (Hyra et al., 2013) and foreclosure concentration (Rugh & Massey, 2010). Thus, future cross-city neighborhood comparisons must more precisely account for, and control, variations in other metropolitan circumstances, such as segregation, that could be influencing the community-level change processes related to the Great Recession.[10]

Further studies should also investigate how federal interventions during the recession and post-recession relate to variations in community-level property values. In this study national interventions, such as Home Affordable Modification Program, Home Affordable Refinance Program and the NSP, which were designed to aid individuals and areas suffering from loan defaults and foreclosures (Immergluck, 2013; Squires & Hyra, 2010), were not addressed. Loan modifications to help troubled borrowers are unevenly distributed across parts of the country and in high-risk versus low-risk communities (Ding, 2013). Thus, in a cross-city neighborhood comparison, the dose of national interventions attempting to stabilize vulnerable homeowners or communities need to be controlled to better isolate the effect of new homebuyer entry on community property value recovery.

Lastly, this study did not assess the impact of the pre-recession, recession and post-recession periods on existing long-term residents. For instance, were there variations on residential, political and cultural displacement in these neighborhoods over the 12-year study period? Did Bronzeville's slower rate of property value escalation, compared to Harlem and Shaw/U Street, preserve the existing character of the neighborhood (see Davidson, 2009; Hyra, 2015)? While this study contributed to a better understanding of neighborhood change dynamics associated with the Great Recession, an exploration of how these forces differentially influenced the lives of long-term residents would have made this a more comprehensive comparative study.

The Great Recession disrupted millions of lives and nearly brought down the entire US financial system and while many studies have investigated the institutional causes and consequences of this monumental economic calamity, few have explored its long-term impact on urban African American communities. Black communities were disproportionately, compared to White areas, the targets of unsustainable financial home lending products which devastated many of these areas. However, during the post-recession recover period some minority communities have recovered, while others remain dotted with foreclosures. This study suggests that Great Recession-related dynamics influence property value trajectories of urban African American neighborhoods, yet race and class neighborhood transitions as well as citywide dynamics mediate the effects of national economic decline and recovery.

Notes

1. One indicator of a subprime loan is if it was a first-mortgage loan originated at 300 basis points, or 3%, above the going prime rate or 500 basis points, or 5%, above the prime rate for a second lien mortgage. In 2004, the federal government started tracking high cost loan originations. We use subprime and high cost loans interchangeably.
2. Greenlining refers to the influx of high-priced mortgage credit into previously redlined underserved neighborhoods.
3. With increases in subprime lending, between 1994 and 2005, the African American homeownership rate increased from 42% to 49% (Gramlich, 2007).
4. Rather than leading to neighborhood decline, some studies suggest that in certain circumstances foreclosures can spur a pattern of neighborhood reinvestment. In certain neighborhoods the availability of below market properties, brought on, in part, by foreclosure concentration, might encourage investors to buy homes, renovate them and sell them to newcomers who perceive that the neighborhood properties are good values compared to other homes in more expensive parts of the city (Li & Morrow-Jones, 2010; Maeckelbergh, 2012).
5. We assess the tightness of metropolitan housing market conditions through comparing the homeownership vacancy rates in the Chicago, New York and Washington, DC MSAs. We assume that upper- and middle-income people are more willing to move to and invest in low-income neighborhoods in metropolitan areas with tighter housing markets (see Aalbers, 2011; Guerrieri, Hartley, & Hurst, 2010). Metros with lower vacancy rates are considered tighter housing markets.
6. The data presented on these communities are based on specific boundaries. Bronzeville consists of Chicago's Douglas and Grand Boulevard districts. Harlem's geography refers to Central Harlem, which is New York City's Manhattan Community District 10. Shaw/U Street's boundaries are 15th Street on the west, Florida Avenue on the north, North Capitol Street on the east, and M Street to the south in Northwest, Washington, DC.
7. In 2004, the federal government started tracking high cost loan originations, thus we present high cost lending figures during and after this year.
8. Census data suggest that Harlem's general population was changing as well as. Between 2000 and 2010, the Black population decreased by 18%.
9. We focus the citywide unemployment rate and not the community level rate since these were gentrifying communities and people were moving to these areas. Additionally, we assessed the metropolitan housing market conditions since we assume that upper- and middle-income people would be more willing to move to and invest in low-income neighborhoods in metropolitan areas with tighter housing markets (see Aalbers, 2011; Guerrieri et al., 2010).
10. For instance, Chicago, New York and Washington, DC have different levels of segregation and this might have influenced, along with other metropolitan factors, Bronzeville, Harlem and Shaw/U Street's susceptibility to subprime loans and associated foreclosures. In 2000, the Black/White dissimilarity index for Chicago, New York and Washington, DC's MSA was 80.4, 79.5 and 63.0, respectively, meaning that Chicago was the most segregated.

Disclosure statement

No potential conflict of interest was reported by the authors.

References

Aalbers, Manuel B. (2011). *Place, exclusion, and mortgage markets*. Malden, MA: Wiley-Blackwell.

Abbott, Carl. (1999). *Political terrain: Washington, D.C. from Tidewater town to global metropolis*. Chapel Hill: The University of North Carolina Press.

Anacker, Katrin B., Carr, James H., & Pradhan, Archana. (2012). Analyzing foreclosures among high-income Black/African American and Hispanic/Latino borrowers in Prince George's County, Maryland. *Housing and Society, 39*(1), 1–28.

Anenberg, Eliot, & Kung, Edward. (2012). *Estimates of the size and source of price declines due to nearby foreclosures: Evidence from San Francisco*. Washington, DC: Federal Reserve Board of Governors.

Avery, Robert B., Brevoort, Kenneth P., & Canner, Glen B. (2006). The 2005 HMDA data. *Federal Reserve Bulletin, 8*, 123–166.

Avery, Robert B., Brevoort, Kenneth P., & Canner, Glen B. (2007). The 2006 HMDA data. *Federal Reserve Bulletin, 93*, 73–109.

Avery, Robert B., & Brevoot, Kenneth P. (2011). *The subprime crisis: Is the Government housing policy to blame?* Finance and Economics Discussion Series. Washington, DC: Division of Research and Statistics, Board of Governors of the Federal Reserve System.

Been, Vicki, Ellen, Ingrid, & Madar, Josiah. (2009). The high-cost of segregation: Exploring racial disparities in high-cost lending. *Fordham Urban Law Journal, 36*, 361–384.

Bennett, Larry. (2010). *The third city: Chicago and American urbanism*. Chicago: The University of Chicago Press.

Bostic, Raphael W., Engel, Kathleen C., McCoy, Patricia A., Pennington-Cross, Anthony, & Wachter, Susan. (2008). State and local anti-predatory lending laws: The effect of legal enforcement mechanisms. *Journal of Economics & Business, 60*(1–2), 47–66.

Boyd, Michelle R. (2008). *Jim Crow nostalgia: Reconstructing race in Bronzeville*. Minneapolis, MN: University of Minnesota Press.

Bunce, Harold L., Gruenstein, Debbie, Herbert, Christopher, & Scheessele, Randall M. (2001). *Subprime foreclosures: The smoking gun of predatory lending?* Washington, DC: US Department of Housing and Urban Development.

Calem, Paul S., Hershaff, Jonathan E., & Wachter, Susan M. (2004). Neighborhood patterns of subprime lending: Evidence from disparate cities. *Housing Policy Debate, 15*(3), 603–622.

Center on Budget and Policy Priorities. (2013). *Chart book: The legacy of the Great Recession*. Washington, DC: Author.

Clark, Kenneth, B. (1965). *Dark ghetto. Dilemmas of social power*. New York: Harper and Row Publishers.

Coleman, Candace. (2012). *Gentrification in the wake of the subprime mortgage crisis* (Master's thesis). Sanford School of Public Policy, Duke University, Durham, NC.

Collins, Glenn. (2010, September 7). Marcus Samuelsson opens in Harlem. *New York Times*.

Davidson, Mark. (2009). Displacement, space and dwelling: Placing gentrification debate. *Ethics, Place & Environment, 12*(2), 219–234.

Ding, Lei. (2013). Servicer and spatial heterogeneity of loss mitigation practices in soft housing markets. *Housing Policy Debate, 23*(3), 521–542.

Drake, St Clair, & Cayton, Horace R. (1945). *Black metropolis: A study of Negro life in a northern city*. Chicago: The University of Chicago Press.

Dreier, Peter, Bhatti, Saqib, Call, Rob, Schwartz, Alex, & Squires, Gregory. (2014). *Underwater America: How the so-called housing "recovery" is bypassing many American communities*. Berkeley, CA: Haas Institute for a Fair and Inclusive Society.

Ellen, Ingrid, Lacoe, Johanna, & Sharygin, Claudia A. (2013). Do foreclosures cause crime? *Journal of Urban Economics, 74,* 59–70.

Engel, Kathleen C., & McCoy, Patricia A. (2011). *The subprime virus: Reckless credit, regulatory failure, and next steps.* New York: Oxford University Press.

Federal Housing Finance Agency. (2013, July 23). FHFA house price index up 0.7 percent in May. *FHFA News Release.*

Freeman, Lance. (2006). *There goes the hood.* Philadelphia, PA: Temple University Press.

French, George E. (2009). A year in bank supervision: 2008 and a few of its lessons. *Supervisory Insights, 6*(1), 3–18.

Gerardi, Kristopher, Rosenblatt, Eric, Willen, Paul S., & Yao, Vincent W. (2012). *Foreclosure externalities: Some new evidence* (Working Paper No. 2012-11). Atlanta, GA: Federal Reserve Bank of Atlanta.

Gerardi, Kristopher S., Foote, Christopher L., & Willen, Paul S. (2011). Reasonable people did disagree: Optimism and pessimism about the U.S. market before the crash. In Susan M. Wachter & Marvin M. Smith (Eds.), *The American mortgage system: Crisis and reform* (pp. 26–59). Philadelphia: University of Pennsylvania Press.

Gillette, Howard. (1995). *Between justice & beauty: Race, planning, and the failure of urban policy in Washington, D.C.* Philadelphia: University of Pennsylvania Press.

Goetz, Edward G. (2013). *New Deal ruins: Race, economic justice, and public housing policy.* Ithaca, NY: Cornell University Press.

Gotham, Kevin F. (2009). Creating liquidity out of spatial fixity: The secondary circuit of capital and the subprime mortgage crisis. *International Journal of Urban and Regional Research, 33* (2), 355–371.

Gramlich, Edward M. (2007). *Subprime mortgages: America's latest boom and bust.* Washington, DC: The Urban Institute Press.

Grossman, James R. (1989). *Land of hope: Chicago, Black southerners, and the great migration.* Chicago: The University of Chicago Press.

Guerrieri, Veronica, Hartley, Daniel, & Hurst, Erik. (2010). *Endogenous gentrification and housing price dynamics* (Working Paper No. #16237). Cambridge, MA: National Bureau of Economic Research.

Halpern, Robert. (1995). *Rebuilding the inner city: A history of neighborhood initiatives to address poverty in the United States.* New York: Columbia University Press.

Hirsch, Arnold R. (1998). *Making the second ghetto: Race & housing in Chicago, 1940-1960.* Chicago: The University of Chicago Press.

Hwang, Jackelyn, & Sampson, Robert. (2014). Divergent pathways of gentrification: Racial inequality and the social order of renewal in Chicago neighborhoods. *American Sociological Review, 79*(4), 726–751.

Hyra, Derek S. (2008). *The new urban renewal: The economic transformation of Harlem and Bronzeville.* Chicago: The University of Chicago Press.

Hyra, Derek S. (2012). Conceptualizing the new urban renewal: Comparing the past to the present. *Urban Affairs Review, 48*(4), 498–527.

Hyra, Derek. (2015). The Back-to-the-city movement: Neighhourhood redevelopment and processes of political and cultural displacement. *Urban Studies, 52*(10), 1753–1773.

Hyra, Derek S., Squires, Gregory D., Renner, Robert N., & Kirk, David S. (2013). Metropolitan segregation and subprime lending. *Housing Policy Debate, 23*(1), 177–198.

Immergluck, Dan. (2009). *Foreclosed: High-risk lending, deregulation, and the undermining of America's mortgage market.* Ithaca, NY: Cornell University Press.

Immergluck, Dan. (2010). Neighborhoods in the wake of the debacle: Intrametropolitan patterns of foreclosed properties. *Urban Affairs Review, 46*(1), 3–36.

Immergluck, Dan. (2013). Too little, too late, and too timid: The federal response to the foreclosure crisis at the five-year mark. *Housing Policy Debate, 23*(1), 199–232.

Immergluck, Dan, & Smith, Geoff. (2005). Measuring the effects of subprime lending on neighborhood foreclosures: Evidence from Chicago. *Urban Affairs Review, 40*(3), 362–389.

Immergluck, Dan, & Smith, Geoff. (2006a). The external costs of foreclosure: The impact of single-family mortgage foreclosures on property values. *Housing Policy Debate, 17*(1), 57–79.

Immergluck, Dan, & Smith, Geoff. (2006b). The impact of single-family mortgage foreclosures on neighborhood crime. *Housing Studies, 21*(6), 851–866.

Joint Economic Committee. (2007). *The subprime lending crisis: The economic impact on wealth, property values and tax revenues, and how we got here.* Washington, DC: Joint Economic Committee, US Congress.

Keeley, Brian, & Love, Patrick. (2010). *From crisis to recovery: The causes, course and consequences of the Great Recession.* Paris, France: OECD Insights.

Kirk, David & Hyra, Derek. (2012). Home foreclosures and community crime: Causal or spurious association? *Social Science Quarterly, 93*(3), 648–670.

Kuebler, Meghan, & Rugh, Jacob S. (2013). New evidence on racial and ethnic disparities in homeownership in the United States from 2001 to 2010. *Social Science Research, 42*(5), 1357–1374.

Lacy, Karyn. (2007). *Blue-chip black: Race, class, and status in the new black middle class.* Berkeley: University of California Press.

Lacy, Karyn. (2012). All's fair? The foreclosure crisis and middle-class black (in)stability. *American Behavioral Scientists, 56*(11), 1565–1580.

Levitin, Adam J., & Wachter, Susan M. (2013). Why housing?. *Housing Policy Debate, 23*(1), 5–27.

Li, Yanmei, & Morrow-Jones, Hazel A. (2010). The impact of residential mortgage foreclosure on neighborhood change and succession. *Journal of Planning Education and Research, 30*(1), 22–39.

Liebow, Elliot. (1967). *Tally's corner: A study of Negro streetcorner men.* Boston, MA: Little, Brown and Company.

Lin, Zhenguo, Rosenblatt, Eric, & Yao, Vincent W. (2009). Spillover effects of foreclosures on neighborhood property values. *Journal of Real Estate Finance and Economics, 38*(4), 387–407.

Logan, John R., & Molotch, Harvey L. (2007). *Urban fortunes: The political economy of place.* Berkeley: University of California Press.

Maeckelbergh, Marianne. (2012). Mobilizing to stay put: Housing struggles in New York City. *International Journal of Urban and Regional Research, 36*(4), 655–673.

Mahoney, Olivia. (2001). *Douglas/Grand Boulevard: A Chicago neighborhood.* Chicago: Arcadia Publishing.

Massey, Douglas S., & Denton, Nancy A. (1993). *American apartheid: Segregation and the making of the underclass.* Cambridge, MA: Harvard University Press.

Osofsky, Gilbert. (1996). *Harlem: The making of a ghetto: Negro New York, 1890–1930.* Chicago: Ivan R. Dee, Inc.

Pattillo, Mary. (2007). *Black on the block.* Chicago: The University of Chicago Press.

Peterson, Paul E. (1981). *City Limits.* Chicago: The University of Chicago Press.

Quercia, Roberto G., Stegman, Michael A., & Davis, Walter R. (2007). The impacts of predatory loan terms on subprime foreclosures: The special case of prepayment penalties and balloon payments. *Housing Policy Debate, 18*(2), 311–346.

Ranney, David. (2003). *Global decisions/local collisions.* Philadelphia, PA: Temple University Press.

Robinson, Eugene. (2010). *Disintegration: The splintering of Black America.* New York: Random House.

Ruble, Blair A. (2010). *Washington's U Street: A biography.* Washington, DC/Baltimore, MD: Woodrow Wilson Center and Johns Hopkins University Presses.

Rugh, Jacob S., & Massey, Douglas S. (2010). Racial segregation and the American foreclosure crisis. *American Sociological Review, 75*(5), 629–651.

Rugh, Jacob S., Albright, Len, and Massey, Douglas S. (2015). Race, space, and cumulative disadvantage: A case study of the subprime lending collapse. *Social Problems, 62*(2), 186–218.

Sampson, Robert J. (2012). *Great American city: Chicago and the enduring neighborhood effect.* Chicago: The University of Chicago Press.

Sampson, Robert J., Raudenbush, Stephen, & Earls, Felton. (1997). Neighborhoods and violent crime: A multilevel study of collective efficacy. *Science, 277*(5328), 918–924.

Sassen, Saskia. (2012). *Cities in the world economy.* Thousand Oaks, CA: Sage Publications.

Shapiro, Thomas M. (2004). *The hidden cost of being African American: How wealth perpetuates inequality.* New York: Oxford University Press.

Shapiro, Thomas, Meschede, Tatjana, & Osoro, Sam (2013). *The roots of the widening racial wealth gap: Explaining the Black-White economic divide.* Waltham, MA: Institute on Assets & Social Policy.

Shin, Annys. (2013, July 21). Gentrification in overdrive on 14th Street. *Washington Post.*

Sites, William. (2003). *Remaking New York: Primitive globalization and the politics of urban community.* Minneapolis: University of Minnesota Press.

Smith, Marvin M., & Wachter, Susan M. (2011). Introduction. In Susan M. Wachter & Marvin M. Smith (Eds.), *The American mortgage system: Crisis and reform* (pp. 1–4). Philadelphia: University of Pennsylvania Press.

Spear, Allan H. (1967). *Black Chicago: The making of a Negro ghetto, 1890-1920.* Chicago: The University of Chicago Press.

Squires, Gregory D. (2005). Predatory lending: Redlining in reverse. *Shelterforce Online,* #139, January/February.

Squires, Gregory D. (2008). *Do subprime loans create subprime cities? Surging inequality and the rise of predatory lending.* Washington, DC: Economic Policy Institute.

Squires, Gregory D., & Hyra, Derek S. (2010). Foreclosures – Yesterday, today and tomorrow. *City & Community,* 9(1), 50–60.

Stone, Clarence N. (1989). *Regime politics: Governing Atlanta, 1946–1988.* Lawrence: University Press of Kansas.

Taub, Richard P., Taylor, D. Garth, & Dunham, Jan D. (1987). *Paths of neighborhood change: Race and crime in America.* Chicago: The University of Chicago Press.

Taylor, Monique M. (2002). *Harlem: Between heaven and hell.* Minneapolis: University of Minnesota Press.

Taylor, Paul, Kochhar, Rakesh, Fry, Richard, Velasco, Gabriel, & Motel, Seth. (2011). *Wealth gaps rise to record highs between Whites, Blacks and Hispanics.* Washington, DC: Pew Research Center.

Timiraos, Nick. (2013, April 3). Fannie's windfall blurs debate over its fate. *Wall Street Journal.*

US Bureau of Labor Statistics. (2014). *Unemployment rate historical data.* Washington, DC: Author.

US Census Bureau. (2014). *Homeownership vacancy rates for the 75 largest metropolitan statistical areas: 2005–present.* Washington, DC: Author.

US Conference of Mayors. (2007). *US metro economies: The mortgage crisis.* Washington, DC: Author.

US Department of Housing & Urban Development Neighborhood Stabilization Program, Round 1 [HUD NSP1]. (2008). *Neighborhood Level Foreclosure Data* [Machine readable data file]. Retrieved August 14, 2015, from http://www.huduser.gov/portal/datasets/nsp_foreclosure_data.html

US Department of Housing & Urban Development Neighborhood Stabilization Program, Round 3 [HUD NSP3]. (2010). *Data Downloadable by State for Each Block Group* [Machine readable data file]. Retrieved August 14, 2015, from http://www.huduser.gov/portal/datasets/nsp.html

Vale, Lawrence, J. (2013). *Purging the poorest: Public housing and the design politics of twice-cleared communities.* Chicago: The University of Chicago Press.

Venkatesh, Sudhir A. (2000). *American project.* Chicago: University of Chicago Press.

Wial, Howard. (2013). *Metropolitan economies in the Great Recession and after.* Paper presented at the Allied Social Science Associations/Labor and Employment Relations Association Annual Meeting, San Diego, CA.

Wessel, David. (2013, August 30). Best restaurants in Washington, DC's 14th Street corridor. *Wall Street Journal.*

Whelan, Robbie. (2013, May 24). New homes hit record as builders cap supply. *Wall Street Journal.*

Williams, Paul K. (2002). *Greater U Street.* Charleston, SC: Arcadia Publishing.

Williams, Richard, Nesiba, Reynold, & McConnell, Eileen Diaz. (2005). The changing face of inequality in home mortgage lending. *Social Problems,* 52(2), 181–208.

Wilson, William J. (1996). *When work disappears: The world of the new urban poor.* New York: Knopf.

Wyly, Elvin K., Atia, Mona, & Hammel, Daniel J. (2004). Has mortgage capital found an inner-city spatial fix? *Housing Policy Debate*, *15*(3), 623–685.

Wyly, Elvin, Moos, Markus, Hammel, Daniel, & Kabahizi, Emanuel. (2009). Cartographies of race and class: Mapping the class-monopoly rents of American subprime mortgage capital. *International Journal of Urban and Regional Research*, *33*(2), 332–354.

Yin, Robert Y. (2013). *Case study research: Design and methods*. Thousand Oaks, CA: Sage Publications.

Appendix

Table A1. Population percent black.

	1990	2000	2010	% change 2000–2010
Bronzeville	95	92	84	−9
Harlem	88	77	63	−18
Shaw/U Street	67	52	30	−42

Source: US Census.

Table A2. Number of home loans by year.

	2000	2002	2004	2006	2008	2010	2012
Bronzeville	814	1,242	1,494	2,049	658	384	171
Harlem	394	515	917	733	714	641	607
Shaw/U Street	972	1,982	2,363	2,262	1,334	1,758	2,312

Source: HMDA Loan Application Register. Note: Includes all lien positions.

Table A3. Total home lending and high cost (HC) loan amounts (in millions) by year.

	2004 Total	2004 HC	2006 Total	2006 HC	2008 Total	2008 HC	2010 Total	2010 HC	2012 Total	2012 HC
Bronzeville	275	62	403	192	167	29	89	3	45	2
Harlem	372	23	573	69	349	12	273	3	426	1
Shaw/U Street	698	26	889	129	528	11	792	5	996	8

Source: HMDA Loan Application Register.

Table A4. Percent of high cost loan originations by year.

	2004	2006	2008	2010	2012
Bronzeville	23	48	17	3	4.0
Harlem	6	12	3	1	0.3
Shaw/U Street	4	15	2	1	1.0

Source: HMDA Loan Application Register.

Table A5. Median borrower income by year ($,000).

	2004	2006	2008	2010	2012
Bronzeville	73	86	84	79	79
Harlem	85	139	128	105	144
Shaw/U Street	103	123	120	135	144

Source: HMDA Loan Application Register.

Table A6. Home loan borrower annual income by community and year (values in %).

	2000	2002	2004	2006	2008	2010	2012
Bronzeville							
$0–$50k	–	22	16	10	11	13	17
$51–$100k	–	53	58	53	54	61	54
$101–$200k	–	22	23	31	29	21	26
Over $200k	–	3	3	6	6	5	3
Harlem							
$0–$50k	30	13	16	6	9	6	5
$51–$100k	39	37	45	29	29	43	25
$101–$200k	27	39	28	33	35	32	40
Over $200k	4	11	11	32	27	20	31
Shaw/U Street							
$0–$50k	–	–	10	5	5	2	2
$51–$100k	–	–	39	29	33	25	23
$101–$200k	–	–	40	47	40	49	49
Over $200k	–	–	12	19	22	24	26

Source: HMDA Loan Application Register (excludes missing income).

Table A7. Total population.

	1990	2000	2010	% change 2000–2010
Bronzeville	66,549	54,476	40,167	−26
Harlem	99,519	107,109	115,723	+8
Shaw/U Street	29,567	29,741	34,750	+17

Source: US Census.

Neighborhood change beyond clear storylines: what can assemblage and complexity theories contribute to understandings of seemingly paradoxical neighborhood development?

Katrin Grossmann[a] and Annegret Haase[b]

[a]Faculty of Architecture and Urban Planning, University of Applied Sciences Erfurt, Germany; [b]Department of Urban and Environmental Sociology, Helmholtz-Centre for Environmental Research, UFZ, Germany

ABSTRACT

Neighborhood research today is largely concerned with two central aspects of neighborhood development: gentrification and decline. This paper sets out to enrich the discourse on neighborhood change, especially that concerned with so-called declining neighborhoods, by drawing on assemblage and complexity theories. These approaches emphasize processes, interdependencies, uncertainties, surprising shifts, and feedback loops in the production of specific spatial formations. We apply this framework in an examination of the development of two neighborhoods in Leipzig: an inner-city district and a large housing estate. We identify internal and external factors impacting these neighborhoods' trajectories and demonstrate how various multidirectional shifts are crucial to the specific paths—and the understanding—of these neighborhoods' development. From a conceptual perspective, we advocate for the use of assemblage thinking in addition to existing approaches to neighborhood change.

Introduction: why we hesitate to write about neighborhood decline

This paper sets out to enrich the discourse on neighborhood change, particularly the case of so-called declining neighborhoods. It has been our experience that many neighborhoods in post-socialist states have taken surprising and unpredictable shifts and turns. The two most prominent conceptual explanations offered by neighborhood research, gentrification and neighborhood decline, seem tempting but not entirely applicable to a post-socialist context. In the case of inner-city neighborhoods in many post-socialist cities, some elements of gentrification processes can be identified, but the evidence only partially approximates the conditions generally associated with processes of gentrification. There are not only physical improvements, rent gaps, and an influx of low-income, highly educated households, but also vast vacancies, infrastructural neglect, and an influx of migrant households.

Our concern with using inherited theories to explain these processes is that if we look for decline, we are likely to find decline; if we look for gentrification, we are likely to find gentrification. Therefore, this paper sets out to take a broader explanatory view of the processes observed. We apply a long-term perspective that encompasses crucial turning points in post-socialist neighborhood trajectories, including the phase of political change in the 1980s and 1990s. This allows us to contribute to current discussions in urban research about the limits of styles of theorizing that identify similar phenomena across a variety of cases and explain these phenomena with a common cause.

The literature on assemblage thinking in urban research develops an ontological point of view that claims to overcome reductionist, linear, and causal thinking in favor of a better understanding of constant change, unexpected effects, shifts, and local context. As Anderson and Mc Farlane (2011, p. 126) puts it, assemblage thinking "attend[s] to a lively world of differences." From such a perspective, scholars aim to better explain informal settlement development (McFarlane, 2011), the energy vulnerability of households (Day & Walker, 2013), political support movements (Davies, 2012), policy transfers (McFarlane, 2011), urban planning (e.g., De Roo, Hillier, & Joris, 2012; Portugali, Meyer, & Stolk, 2012), and urban theory in general (Farías & Bender, 2010; Manson & O'Sullivan, 2006; Thrift, 1999).

In this paper, we examine the development of an inner-city district and large housing estate in Leipzig in order to test whether assemblage ontologies can enrich neighborhood change research in a post-socialist context. The inner-city district—Leipzig's inner east—serves well as a site for both decline and gentrification studies. After years of economic decline, abandonment, and decay, it has recently experienced significant rejuvenation through the influx of young households from the region as well as through foreign immigration. The large housing estate—Leipzig-Grünau—lost half of its population after 1990 but remained surprisingly stable in terms of the socioeconomic status of its residents (Kabisch & Großmann, 2013). Recently, the estate has seen a population increase, but in-migration is bringing more households with lower socioeconomic status. This neighborhood could thus serve as a research-typical site for an analysis of neighborhood decline.

The structure of this paper reflects our primary interest in better understanding the development trajectory of these neighborhoods rather than explaining theory development solely on the basis of inherited theories of gentrification and decline. Instead of a theoretical introduction, we first describe the development in the two neighborhoods from a long-term perspective. Next, we outline what the neighborhood decline or gentrification frameworks would tell us about the cases and point to remaining open questions. We then discuss the cases using complexity and assemblage thinking. In the conclusion, we consider how—and to what extent—various approaches to theorizing neighborhood change can be interlinked in fruitful ways. From a conceptual perspective, it is the purpose of this paper to argue for the merits of assemblage theory in producing a more comprehensive picture of neighborhood change. This paper explores how assemblage and complexity thinking are helpful for looking at neighborhood evidence from a context-sensitive and relational perspective. We argue for the use of assemblage thinking in addition to existing approaches, not as a replacement.

Along with municipal statistical data, we draw from our long-term engagement with the districts in this study, including our own survey data, household and expert interviews, and participant observations. In the case of Leipzig-Grünau, we have unique longitudinal survey data on the social structures, perceptions, and housing satisfaction of residents, beginning in 1979, from Kabisch & Grossmann (2013). Additional information is drawn from analyses of the governance of urban shrinkage in Grünau, where expert interviews were conducted with representatives of the city administration, neighborhood management and initiatives, and housing companies.

Leipzig, a regrowing city, and its neighborhoods

Leipzig has experienced turbulence in its population numbers throughout the last decades (Figure 1). Following a sharp decline in population, the city has recently experienced a steep increase in population levels, last seen at the beginning of the twentieth century.

As in many eastern German cities, post-1989 political changes and deindustrialization led to a steep rise in unemployment rates and a wave of job-related out-migration. In addition, the second half of the 1990s brought a wave of out-migration to the suburbs (see also Rink, Haase, Grossmann, Couch, & Cocks, 2012). Out-migration

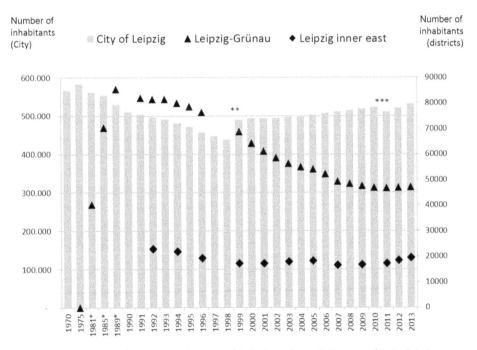

Figure 1. Population development in the city of Leipzig, Leipzig Grünau, and Leipzig's inner east. Note: The estate Leipzig-Grünau comprises the current administrative districts of Grünau-Ost, Schönau, Grünau-Mitte, Grünau-Nord, Grünau-Siedlung, and Lausen-Grünau. Leipzig's inner east includes the districts of Neustadt-Neuschönefeld and Volkmarsdorf (see Table 1). Whereas the numbers of the left axis show data for the city of Leipzig, those on the right axis show data for the two neighborhoods.
Source: Authors' work based on municipal data.

led to an overall population decline, a loss of roughly a fifth of the city's population by the end of the 1990s, and relatively rapid aging of the population that remained. Then, from 1998 onwards, Leipzig experienced an increased influx of young in-migrants, mainly from the surrounding region. This influx has led to a stabilization of the city's population and a rejuvenation of the inner city districts. The last 3 years have brought even more intense in-migration, attracting, in particular, young people from all over Germany and some foreign in-migration. Most districts of the city are now gaining population.

Throughout this time, the neighborhoods developed in an uneven fashion, with hotspots of population decline and rapid reurbanization in close proximity to each other. Neighborhood pathways derived from a number of overlapping internal and external trends. Before 1989, the inner city areas had entered a state of dilapidation; investments initiated by the state were used for the construction of new housing estates on the urban fringe or the demolition of inner city areas. After the German reunification, the inner city saw steady and large-scale investment induced by urban development programs, strategic planning, and tax incentives for investors. The large housing estates also saw improvements in housing quality, infrastructure, and the design of open spaces.

Out-migration impacted all types of neighborhoods in the 1990s, but in-migration in the 2000s was concentrated in inner city neighborhoods. Suburbanization added new settlement structures to the urban fringe, most of which were incorporated into the administrative boundaries of the city.

Case study: Leipzig-Grünau

Leipzig-Grünau is a large housing estate situated in the western outskirts of the city. During its 37-year existence, it has experienced numerous shifts and turns. We will describe these in a chronological fashion with an emphasis on their interdependencies.

Construction of Leipzig-Grünau began in 1976 and continued through the 1980s. Buildings were erected both by the state-owned housing company and by several local cooperatives. The city's housing administration and/or cooperatives assigned residents to housing according to their employment type; for example, the "transport" cooperative gave flats to railway workers, and the "Unitas" cooperative provided flats for university employees. A mix in the socioeconomic status of residents was explicitly planned; young families were a priority target group for the housing estate. By 1989, the estate had a population of approximately 85,000 inhabitants.

In relation to other segments of the (state-controlled) housing market at that time, Grünau was a privileged location, with flats of a relatively high housing quality situated within a fresh air corridor close to a lake that forms a natural recreation area. Towards the end of the 1980s, average satisfaction with the estate started to decrease (Figure 2), which can be explained by a combination of internal and external factors. The higher density at the youngest parts of the estate was subject to criticism, as was the area's uniform development.

With the political changes in 1989, the context of the estate changed in many ways. The societal appreciation of "socialist" housing estates declined significantly, reflected also in a steep drop in Grünau's population satisfaction. Out-migration began, but new

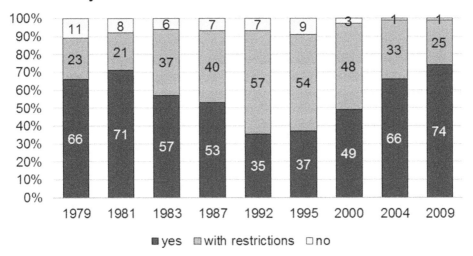

Figure 2. Rates of people feeling comfortable in Grünau, 1979–2009.
Source: Authors' own calculations, based on surveys (1979–2009).

residents arrived swiftly and filled the gaps, because housing shortages were still omnipresent. Out-migration together with the aging of residents altered Grünau's demographic characteristics completely. Within just 20 years, the overrepresentation of young families disappeared. Now, older residents are overrepresented.

In 1995, the housing market was liberalized, the state housing company became municipal, cooperatives remained in place, and private housing started to develop. Although the overall population of the city decreased sharply, refurbishments and new construction added new housing in the inner city and suburbs to the market. As a consequence, the relative attractiveness of Grünau decreased. Old built-up structures were now highly valued because they provided historical and individual housing environments, whereas housing estates like Grünau were labeled as "gray," "uniform", and of "low urban quality." The newly accessible possibility of building one's own single-family house also led to a relative loss of attractiveness for housing estates. As a result, Grünau has struggled with a negative external image since the 1990s.

The 1990s were a period of constant investment. Housing structures were modernized, with improvements to indoor comfort for many residents and new facades changing gray concrete to pastel walls. The city, using state and federal funding schemes, invested in both infrastructure and open space improvements. A public swimming pool was built. Public green spaces and central open spaces were subject to design competitions, which also brought in new public art. Private investment created a large shopping mall and cinema in the center of the estate. Consequently, inside the district, housing satisfaction grew again.

At the end of the 1990s, however, housing vacancies became evident, and a new phase of demolitions and de-densification began, including school closures and infrastructure loss. Population numbers fell from a peak of approximately 85,000 inhabitants in 1989 to fewer than 43,000 in 2011 (City of Leipzig, 2012b). As a consequence, housing vacancy increased to 26% in Leipzig-Grünau in 2003 (vs. 19% in the city of Leipzig, City of Leipzig, 2012a). To counteract the high vacancy rate, more than 6,800 flats in the estate were demolished after 2000 (City of Leipzig, 2010). The district manager estimated in an interview in 2013 that approximately a third of all schools and kindergartens had been closed. The policies that precipitated the demolitions were introduced on a national level with the urban restructuring program "Stadtumbau Ost." In these policies, integrated planning for a reduced number of people was made a pre-condition for demolition subsidies. In this integrated plan for the city as a whole (City of Leipzig, 2000), Grünau was defined as a location for concentrated demolition.

Though there were expectations of a decline in average socioeconomic status due to selective mobility patterns—as typically described by the neighborhood decline litera-ture—survey data do not support such a trend. Despite the heavy out-migration, the available social status indicators do not show a significant drop in socioeconomic status.

The share of respondents with higher education degrees decreased slightly only in the early 1990s (Figure 3). This drop was mainly caused by a relative decline in respondents with technical college education. The share of respondents with higher education returned to the levels of the 1970s and early 1980s and remained stable over the following years. Because the share of tertiary education is rising in society as a whole, this actually indicates a gradual, relative decline in educational status. Still, the waves of out-migration did not decrease the area's average levels of education. Newcomers have, on average, a slightly lower educational status, but this did not contribute to an abrupt decline in educational status because the overall population influx was rather low.

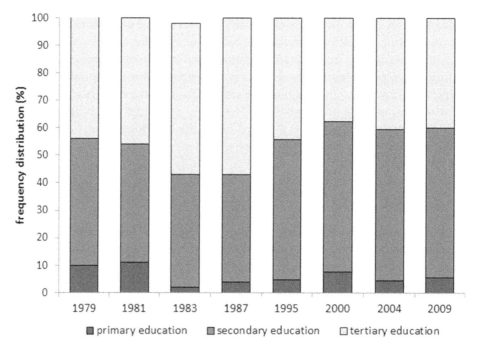

Figure 3. Educational level of residents in Leipzig-Grünau, 1979–2009.
Source: Authors' own calculations, based on surveys (1979–2009).

In parallel with the growing number of residential vacancies, ownership structures changed. The municipal government was the trailblazer for demolitions and also sold large portions of its properties to private companies or hedge funds that saw a possibility of profiting from this housing segment. Today, the municipal government is basically withdrawn from the outer parts of the estate. The cooperatives have sold and demolished their buildings to a lesser extent. Private companies' strategies range from disinvestment to a strong engagement with the district's improvement efforts and sponsoring activities. Interestingly, the oversupply on the market turned housing companies into innovators for attracting residents: they offered gifts to people willing to move in, redesigned flats individually, adapted flats to meet the needs of elderly residents, constructed playgrounds close to buildings, and even introduced a bakery service on Sundays. Demolitions were at first experienced as a shock and a crisis by residents, which led to protests and conflicts. Some years later, changes such as lower housing densities and increased green spaces were appreciated by the majority of residents (Kabisch and Großmann, 2010, pp. 69–76).

The demographic crisis that Grünau experienced has also contributed to a strong network of actors that still exists today. Under the moderation of district management, housing companies, civic groups, economic actors, and the city administration have formed a strong network that fosters various activities to stabilize the estate. They promote strategic investment in amenities and infrastructure that attract new inhabitants and prevent further out-migration.

Recently, the estate has seen new in-migration and—for the first time in 20 years—a net gain in residents (City of Leipzig, 2014). This trend is primarily related to the recent steep gain in population throughout the city, caused by both job growth and an improvement in the city's public image, leading to nicknames such as "Hypezig." What this means for the future is unclear. It is very likely that demographic decline will slow or halt. Vacancies will also decline; even a new multifamily house is being realized on former demolition plots by one of the cooperatives. Grünau today belongs to the districts with lower housing prices, both because flats are smaller than those in many other districts and prices per square meter are comparatively low. In as early as 2011, housing companies observed that low-income households were relocating due to rising prices in the inner city districts and that Grünau might—demographically—profit from dislocations in the inner city, reducing Grünau's vacancies. Over the last 10 years, in-migrants have been characterized by, on average, a lower social status and different demographics from the long-term residents (Kabisch & Großmann, 2010, pp. 53–56). However, the low number of in-migrants has not affected the overall socioeconomic status indicators significantly. Thus, increasing in-migration could lead to a sudden drop in socioeconomic status, but it does not necessarily have to.

Case study: Leipzig's inner east

The eastern part of Leipzig's inner city represents an old built-up residential area with a predominance of workers' tenement building stock. At present, the area provides housing for approximately 42,000 inhabitants (2008) and has a territory of 340 hectares. Starting in the middle of the nineteenth century, the area developed as a workers' district, eastwards of the Leipzig city center. Until 1945, the area represented a typical workers' district with mixed uses as a residential, retail, and industrial area with a high population density and relatively basic housing standards (Behling & Henn, 2010, pp. 18–19).

During the German Democratic Republic (GDR) era, two important processes occurred in the area: (1) large-scale dilapidation and physical decay and (2) selective renovation and restructuring activities in the late 1970s and the first half of the 1980s, when the old building stock, which comprised entire blocks, was demolished and replaced with pre-fabricated housing. Vacant housing evolved in this area before 1989 —not because of oversupply, as there was a housing shortage in the GDR, but as a consequence of poor maintenance and decay, which made buildings and flats unin- habitable. As for the population, selective out-migration led to an above-average proportion of older people and to below-average proportions of working-age people, women, and children. Furthermore, the settlement of economically disadvantaged households and released criminals in old tenement buildings led to a concentration of very poor and excluded social groups in this area by the end of the GDR era (Richter, 2000, cited in Behling & Henn, 2010, pp. 19–20).

The most massive exodus of population occurred in the first decade after the political turnaround. In many areas, an "exchange" of the population was observed, the rate of which amounted to over 80% of the population living there for certain neighborhoods in the early 1990s (Haase, Kabisch, & Steinführer, 2004). In general, the better-educated and younger residents moved away from the area, and poor and older households were left behind. This trend explains why Leipzig's inner east became a target area for urban restructuring relatively early, in November 1990. With the establishment of three renewal areas, the prerequisite for "fast action," set against the background of decay and decline in the area, was created (Gerkens, Hochtritt, & Seufert, 2010, p. 76). The 1990s were thus characterized by the overall objective of safeguarding and improving the dilapidated urban infrastructure; investment was encouraged through tax incentives.

By the year 2000, when out-migration in the city as a whole brought the issue of housing vacancies onto the urban political agenda, Leipzig's inner east had one of the highest vacancy rates in the city. In a city-wide comprehensive plan for dealing with housing vacancies, a subplan for Leipzig's inner east foresaw the demolition of 25% of the building stock (8,000 flats) as well as the idea of checkering the area with enlarged green areas interspersed with housing (City of Leipzig, 2000).

However, new in-migration had already started in the late 1990s, when the area became interesting for a variety of residents, such as young, small households; students; low-income families, single parents, and migrants. In-migration resulted in both the rejuvenation and diversification of the residential population. Whereas these trends were also typical for other inner-city districts, the increasing share of (international) migrants from many countries, of which the largest groups came from Russia, Ukraine, and Vietnam, was a specific feature of Leipzig's inner east. It developed into Leipzig's first real migrant area, and currently has shares of 18–20% of migrants in some districts (City of Leipzig, 2012c).

In the mid-2000s, there was a political reorientation on issues related to urban restructuring and demolition. With new, increasing in-migration and population regrowth in Leipzig as a whole, the discussion of further demolition ceased. Other contributing factors were a lack of money, the inability to persuade actors such as small, private owners (of single buildings) to become involved, and normative perceptions of the high value of old buildings that led to opposition toward further demolition of pre-war housing stock. Subsequently, projects encouraging community cohesion, the refurbishment of the built environment and public spaces, and small-scale upgrading measures received more attention. The main objective was now to stabilize the area as a residential location and maintain the status quo, that is, no further out-migration, no further decline, and increased social cohesion (STEK LeO, 2011).

In addition to stabilization, several initiatives initiated upgrading during the 2000s. The municipality tried to support the (emergent) reurbanization trends in the area, as mentioned above. Planners' idea was to attract middle-class families to the area, a strategy that did not have much in common with reality; most of the newcomers were low-income households, students, early-stage professionals, and immigrants from other countries. Therefore, in the mid-2000s, political attention also focused on these groups. The municipality even undertook actions to attract more immigrants to the area as residents and shop owners. However, despite its excellent location, public transport connections, and the ready availability of numerous abandoned lots, support schemes for owner-occupied housing were not very successful and remained limited to small parts of the area.

An examination of the sociodemographic characteristics of two districts of Leipzig's inner east, Neustadt-Neuschönefeld and Volkmarsdorf, for the 2000s (Table 1) reveals that the residents are young, considerably younger than the city of Leipzig as a whole. These districts have a relatively high youth rate and a relatively low elderly rate. The high proportion of migrants, which exceeds Leipzig's average by far, could point to an area that would face prejudice in media and governmental imaginations; the positive migration balance and large share of young population could, by contrast, point to an area that promises pronounced population growth, also through families, in combination with the high youth rate and general rejuvenation. The large share of one-person households points to two different issues—an increasing share of older, one-person households (due to general aging) and a rising influx of young, one-person households (as a part of reorganization, since approximately 2000; see Haase et al., 2010). In sum,

Table 1. Sociodemographic characteristics of two districts of Leipzig's inner east (Neustadt-Neuschönefeld—NN; Volkmarsdorf—VM) 2003, 2007, 2011.

	Mean age (years)	Share of inhabitants <40 years (%)	Youth rate (%)	Elderly rate (%)	Share of migrants (%)	Share of one-person households (%)
NN 2003	37.3	60.0	15.2	16.5	14.8	n.a.
NN 2007	38.0	57.0	18.5	19.7	20.0	59.0
NN 2011	37.2	60.6	18.3	17.9	20.4	57.6
VM 2003	38.1	56.0	16.1	17.2	13.3	n.a.
VM 2007	38.8	54.0	18.5	19.9	18.0	59.0
VM 2011	39.5	54.5	19.0	21.0	19.0	58.2
Leipzig 2003	42.9	47.8	13.5	26.4	5.2	n.a.
Leipzig 2007	44.0	44.8	15.1	32.6	6.4	51.9
Leipzig 2011	43.8	46.1	18.2	33.7	5.2	50.7

Source: City of Leipzig, Ortsteilkataloge 2004, 2008, 2012.

the numbers do not point clearly in one direction or the other, towards "up" or "down," towards gentrification or decline.

Today, the number of inhabitants in the area is growing rapidly. Compared to other districts of Leipzig, the area had the highest population growth in 2012. The population structure is has diversified since the early 2000s; however, there is still a concentration of low-income households and migrants, so the area suffers from a rather negative image (Haase et al., 2010). Since 2008, cuts in funding have made the previously followed forms of cooperation more difficult. A gradual erosion of the ambitious approach has occurred during the past few years. A strategic meeting and brainstorming session in spring 2011 revealed that money, capacities, and ground-breaking ideas on how to develop the area further were lacking—today's governing reality has become much more similar to "muddling through" than to the consistent restructuring or activation-oriented policies that were applied in the first half of the 2000s. At the same time, the area can now move more quickly towards focusing on upgrading policies than was possible in the past. During the year 2014, set against Leipzig's fast regrowth in the 2010s, vivid in-migration to Leipzig's inner east was observed, with young people, entrepreneurs, and artists realizing housing and urban gardening projects.

Approaching the case study neighborhoods through the Heuristics of Neighborhood Decline and Gentrification

Both our cases can serve as contexts for the application of neighborhood decline research; furthermore, the inner east could be a considered a case of burgeoning gentrification. In what follows, we examine what these heuristics help us to understand in the two case study areas.

The experiences of specific pathways of large post-war housing estates in the United States and Western Europe fostered *models on neighborhood decline*, which identified a range of drivers and mechanisms leading to a loss of socioeconomic status for neighborhoods (Spicker, 1987; Power & Tunstall, 1995; Power, 1997; Prak & Priemus, 1986; Skifter Andersen, 2003; for an overview, see Van Beckhoven, Bolt, & van Kempen, 2009). Researchers developed a variety of approaches. Some blame the modernist design of the estates as the causal factor responsible for increasing social deprivation, decay, and criminality—for example, the concept of "defensible space" (Newman, 1972) or "Utopia on Trial" (Coleman, 1985). In rejecting this argument, authors such as Spicker (1987) maintain that increasing poverty in society in general is the driving force for further spatial concentrations of poverty and deprivation in housing estates. Other authors ask why more affluent households leave while less privileged groups move in or are "trapped" (Musterd & Van Kempen, 2007) and explain this by considering the economic constraints of households in segmented housing markets. Such arguments are characteristic not only of North American research but also of some Western European approaches (see reviews in Skifter-Andersen, 2003; Van Beckhoven et al., 2009). Others are more interested in the mutual interactions of different causes. For example, Power (1997) emphasizes the interaction between social problems and physical decay but not in the deterministic manner of Newman or Coleman. She argues instead that the management of estates is decisive, both in explaining the decline of estates and in developing policies to combat decline.

Other researchers have developed relatively complex models of neighborhood decline that encompass not only design and mobility but also aspects such as the technical state and potential physical decline, or economic stress for landlords and housing companies (Prak & Priemus, 1986; Skifter-Andersen, 2003). While searching for the factors that contribute to neighborhood increases in a lower-income population, Grigsby, Baratz, Galster, and Maclennan (1987) developed a framework that distinguishes a number of external and internal factors, all of which contribute to such a progression of events, starting with housing market forces via the influence of market intermediaries, which lead to particular landlord decisions on (shoddy) maintenance and residential mobility. A characteristic of these rather complex models is that they focus on actors and frameworks. In addition, they include a number of feedback loops that signal how different factors may reinforce each other. The heuristic of a downward spiral is one of the metaphors used here and fits this approach's focus on explaining the decline of post-war housing estates.

The second grand narrative of neighborhood change today is found in the gentrification literature, which examines the upward pathways of neighborhood development, that is, the "upgrading" of neighborhoods by selective (more affluent) in-migration and the displacement of poorer inhabitants. Gentrification approaches either describe a cycle or sequence of upgrading (starting with so-called pioneers entering a dilapidated area and ending up with the conquest of this area by gentrifiers following refurbishment and symbolic upgrading; Clay, 1979) or analyze the process of neighborhood change as an economic process from the perspective of the production of a social process, from the supply side (rent gap, Smith, 1987, 2002), or from the consumption or demand side ("pioneers" and, afterwards, gentrifiers; Hamnett, 2003; Ley, 1994). Within the last years, the focus of many studies has shifted more to the analysis of structural forces, such as underlying real estate and housing market developments, new investment strategies or political/power relations within a neoliberal urban context (Bernt, Haus, & Robischon, 2010; Holm, 2012, 2014).

When looking at Grünau through the lens of neighborhood decline, we can find a number of matching factors. The estate's relatively low position in the local housing market; the concentration of municipal housing; the influence of financialization on the estate, with new private investors coming in to earn quick money; the out-migration of nearly half of the residents; the influx of migrants—are all ingredients of decline analyses. However, why do we simultaneously find enormously high neighborhood satisfaction, especially in the years following population losses and demolitions? Why did public and private investments come in at a time of huge population losses? Why does the share of highly educated residents remain high in a period of rapid out-migration?

For Leipzig's inner east, we can easily show how this neighborhood has gone through 20 years of decline but recently has become likely to turn to a phase of (early-stage) gentrification. The core structural conditions are all there, and the classic groups of early gentrifiers are entering the area right now. However, these conditions have been present for more than 20 years and did not lead to upgrading (or gentrification) after 1990. In the early 2000s, there were even discussions over whether to demolish vacant historic housing stock due to low demand. The first wave of in-migration at the end of

1990s was not one of early gentrifiers moving into a declined but potentially popular area but rather one more mixed in terms of income and household structure. Only very recently did gentrifiers and the attendant real estate interests seem to discover the potential of the area. So why did gentrification or new investment not occur despite the existence of these many gentrification pre-conditions? Why did these things not occur even at a time when Leipzig saw reorganization and rising population numbers?

Cross-referencing the two cases, it becomes—hopefully—obvious why we hesitate to analyze them simply along the above-mentioned lines and according to the logics of clear stories, regardless of whether we refer to them as instances of neighborhood decline or of gentrification. All of the described developments show that neighborhood development in these two areas—and we suppose that they are not just exceptions—is often highly complex and diffused and does not fit into one clear storyline. This leads us to the core endeavor of our paper: examining the development of the areas through the lens of assemblage theory and a complexity approach. We now introduce some core elements of this approach and then introduce more specific aspects while revisiting the neighborhoods.

Inspirations for urban research from assemblage and complexity theory

Complexity theory is an interdisciplinary movement that is increasingly acknowledged in the social sciences; Urry (2005) even speaks of a complexity "turn" in social theory. A single and coherent complexity theory does not exist; rather, a family of theories with similar ontological stances has been developed in an interdisciplinary community that is focused on opposing reductionism and linear causal thinking (e.g., Capra, 1997; Prigogine, 1980). In philosophy, Deleuze and Guattari (1987) outlined what today is recognized as assemblage theory, a mode of reasoning that departs from causal, dichotomic thinking (the rhizome rather than the tree) to embrace the multitude of interconnections that continuously shape and reshape the phenomena that social research aims to understand. DeLanda (2006) further developed this theory to inform understandings of societies, applying it to, among other things, cities and neighborhoods. Assemblages are entities, assembled of component parts, "wholes whose properties emerge from the interaction between parts," characterized by relations of exteriority and by historical processes (DeLanda (2006, p. 5, 10). To think of societal phenomena as assemblages is to look at them as constantly changing configurations of component parts, materialities, infrastructures, varieties of actors and even policies, all of which are interrelated in a multitude of connections and relations. Despite their many differences, approaches organized under the umbrella of assemblage and complexity theories share a systemic perspective. Instead of direct cause–effect relationships, they look at feedback loops, unexpected processes, and new, emerging characteristics.

This approach is being taken up in the publications of social scientists (e.g., Manson & O'Sullivan, 2006; Urry, 2005; Walby, 2009), among them urban researchers and planners (De Roo & Silva, 2010; De Roo et al., 2012; Farías and Bender, 2010; Innes and Booher, 2010; McFarlane, 2011; Portugali et al., 2012; Thrift, 1999), to stimulate new perspectives on a variety of issues. These contributions share an emphasis on the processes and dynamics of the social world that encompass nonlinear developments and feedback loops, and, attendant to this, portray surprise and uncertainty as normal rather than exceptional. Social change from this perspective is the nonreducible outcome of interdependencies and interactions, rather than powerful single drivers or linear causal chains operating similarly in different settings. Rather, context and path dependencies are important frames that explain the different outcomes of similar trends.

Unlike the neighborhood decline or gentrification heuristics, assemblage theory is an ontological frame or strategy (Van Wezemael, 2008) that establishes new priorities and modes of reasoning. Therefore, these theories operate on very different levels. In the categories of theorizing introduced by Beauregard (2012), assemblage and complexity theory are grand theories, establishing "formal arguments … using precisely defined concepts and highly specified relationships … to capture the underlying logic of the reality under investigation" (p. 477), whereas gentrification and neighborhood decline are heuristics "discovering new ideas and operates by pointing to specific aspects of reality or ways of thinking" (481). Thus, the question is not to replace one type of approach with the other but to explore how they might work together to better explain neighborhood change; as DeLanda (2006, p. 119) puts it, assemblage thinking and other concepts can "come together to form a chorus that does not harmonize its different components but interlocks them while respecting their heterogeneity."

Applications to the case study neighborhoods

Approaching our case study neighborhoods from a complexity or assemblage perspective allows a shift in analytical attention. Assemblage theorists are more preoccupied with understanding the components and interactions that drive the development of a neighborhood than they are with a neighborhood's specific developmental trajectory. We will now introduce in more detail the premises of complexity and assemblage thinking most important for our argument, followed by an application of these frameworks to our two case study neighborhoods. We cover four sets of applications: phenomenal emergence, interdependencies, and interactions; policies and governance in an assemblage perspective; the normalcy (as opposed to exceptionalism) of surprise and uncertainty; and processes and dynamics in specific contexts.

Emergence, interdependencies, and interactions

The concept of emergence, which is central to complexity theory, provides us with clues for better understanding the emergence of novel trajectories or a sudden change of direction in neighborhood development. A core argument is that whatever happens in society is rarely due to a single causal mechanism. "Instead, the concrete actualization of events results from the interaction of diverse causal tendencies and counter-tendencies" (Jessop, 1999, p. 3). Change, according to this understanding, derives from ongoing interactions of the various parts of a system and from their interrelationships with other systems and their parts. In this way, new collective properties can emerge.

Assemblage theory is especially useful here, because it highlights the instability of a given entity—which might be a neighborhood, a city or a nation—depending on the scale of analysis (DeLanda, 2006). Thus, emphasis is on the (re)formation of an assemblage as the result of the ever-changing relationships of its component parts. In neighborhoods, not only people, networks, organizations, and buildings but also policies, legal frameworks, and values are such component parts. Further, these component parts are related to other assemblages, either close or far away. A large housing estate is thus related not only to other neighborhoods in the same city but also to other large housing estates, for example, when specific (supra-local) policies are developed to serve a certain type of neighborhood in different cities. Finally, assemblages are themselves part of other assemblages; a historic district is a part of that specific city, but it can also be a part of the assemblage of cultural heritage, as defined by, for example, UNESCO. Accordingly, neighborhoods are products of concrete practices, constructed in relation to other assemblages. Urban research would here try to understand how assemblages are being made or unmade at particular sites of practice (Farías, 2010).

If we apply this perspective to our two case studies, neighborhood changes result not only from changes in composition but also from shifts in relation to other assemblages. Within GDR housing policy, large housing estates are places of priority resource allocation, whereas inner city areas are subject to neglect, decay, and decline. In the aftermath of German reunification, however, the general planning emphasis was on the old building stock, which led to a stigmatization of large housing estates locally and nationally and to upgrading measures in Leipzig's inner east. In effect, Leipzig's inner east became more attractive and became subject to new in-migration at the expense of large housing estates such as Grünau. Thus, there are interrelations between our two case studies when each is conceived as an assemblage; these changes are induced by other, external assemblages. For Leipzig-Grünau, the internal relationships of the component parts of Grünau and the latter's external relations in fact pulled the neighborhood in different directions. As described above, upgrading occurred at the same time as out-migration, which, in DeLanda's terminology, contributed simultaneously to a stabilization and destabilization of the assemblage. In the end, this might explain the area's surprising social stability during politically and demographically turbulent times.

To conclude, the added value of assemblage theory evidenced here is that it focuses on the processes in which component parts are involved. In doing so, the theory helps explain—and this is crucial for the focus of our paper—why the interactions of components can either stabilize or destabilize the assemblage.

Policies and governance from an assemblage perspective

Policy and governance analyses have also been subject to assemblage thinking in recent years (e.g., Allen & Cochrane, 2010; McFarlane, 2011; Van Wezemael, 2008. A main driver of these works is to come to better terms with the processes of policy transfer and the surprising results and failures of such attempts. Van Wezemael (2008, p. 170) shows, for example, that identical strategies can produce very different outcomes, even under similar conditions. Authors writing in this vein aim to highlight the complexity of institutional arrangements, unfolding the "intersecting lines of decision making" (Allen & Cochrane, 2010, p. 1076). Instead of a central state apparatus in possession of far-reaching powers, these researchers posit *powers of reach*, that is, "the ability of the state to permeate everyday life" or the extent to which "actors are drawn within reach" (Allen & Cochrane, 2010, p. 1074). The urban restructuring ("Stadtumbau") program, for example, only influenced owners who were economically dependent on demolition subsidies and who were drawn into networks of demolition planning.

Allen and Cochrane's concept is quite revealing here. Whereas the main urban policy program reacting to the heavy population losses in eastern German cities and emerging housing vacancies (the "Stadtumbau Ost" plan) provided the same instruments for all types of neighborhoods, it had very different outcomes in our two districts of study. In Grünau, the program could access two crucial actors: the owners (the municipal housing company and the larger cooperatives) who owned whole blocks of flats and feared the rising vacancies, and also the city's administration, which supported large-scale demolitions in the outer parts of Grünau. It is the municipality that coordinates most of the activities and decides on the distribution of finances; many other actors depend on these decisions or coordination activities. In contrast, in Leipzig's inner east, owners were seldom willing to demolish houses, which they viewed as their capital. Smaller private owners here were not prepared to demolish housing in central locations with assumed monetary and heritage value. Civic society and mass media supported such perceptions and decisions. Thus, the "Stadtumbau Ost" program had very different effects on the inner east than on Grünau. If the unexpectedly high inflow of new residents to Leipzig's inner east had not arrived, the area might have suffered from further decay. In contrast, Grünau rid itself of large parts of its vacant housing and achieved more open green space.

In sum, assemblage theory shows that power is continually negotiated and renegotiated. It does not simply exist in a top-down fashion. Funding often comes from national funding schemes and thus relates to policy fields and targets defined elsewhere, but the implementation of policies also depends largely on the actors the policy actually reaches. Again, this web of (inter)dependencies is crucial to understanding how policies contribute to neighborhood change.

Normalcy of surprise and uncertainty

If social phenomena stem from interactions and interdependencies between different components (or assemblages) in a nonlinear fashion, then their outcomes are necessarily open. Proponents of the complexity approach therefore emphasize that surprise and uncertainty in outcomes are in fact relatively normal (Cilliers, 1998; De Roo & Silva, 2010). The famous metaphor of the butterfly effect in chaos theory carries this to the extremes by claiming that tiny variations of initial conditions can have dramatic and surprising effects. Complexity theorists argue that the recognizing nonlinear developments, feedback loops, and a reality full of seemingly unexpected events and paradoxes is necessary if scholars hope to understand—and, indeed, influence—today's globalized world (Urry, 2005).

As mentioned above, at first sight, the "neighborhood decline" explanation appears to fit well with both Leipzig's inner east and Leipzig-Grünau. A more detailed analysis, however, reveals that the process does not follow just one logic or trajectory. The decline of Leipzig's inner east must be seen within the framework of the difficult situation facing older, built-up areas in Leipzig until the mid-1990s. After this time period, their trajectories became more varied: some areas became fashionable and were upgraded, whereas Leipzig's inner east was left behind. Urban restructuring and regeneration policies since 2000 have been at best been able to stabilize the area and prevent further decline.

At the same time, the area underwent reorganization as new, heterogeneous population groups arrived. One of the *surprises* for Leipzig's inner east was that the area was the site of Leipzig's first real migrant concentration. The speed of this development, from the mid-1990s onwards, was also surprising. As for the future, there is much *uncertainty* as to the future trajectory of Leipzig's inner east; much will depend on general developments in the city: the in- and outflow of people, its housing market, and the job market situation. Several trajectories are possible, ranging from decline and separation from the development of the city as a whole to continuous upgrading and even gentrification. Leipzig's inner east is a case in point for illustrating what *nonlinear neighborhood development* might look like.

The history of Grünau also reads as a chain of *surprises* and *novelties*. From an intrinsic perspective, the rapidly emerging vacancies at the end of the 1990s came as a surprise. Furthermore, from a neighborhood research perspective, the area's social stability during the time of a population decline is a surprise. The federal program "Urban Restructuring East" (Stadtumbau Ost) was also a novelty in German housing politics, because it fostered the reduction of existing vacancies through the demolition of entire blocks. Thus, within just 20 years, housing blocks were erected and torn down. The national policies touched down and intermixed with the practices of a whole range of local actors, producing conflicts that likely fostered further out-migration due to feelings of insecurity and a fear of general decline. However, this fear quickly turned into acceptance of demolitions. As the 2009 survey shows, most of the remaining residents perceived it (in hindsight) as a necessity, and even framed it as a form of upgrading. Uncertainty accompanied actors' decisions in these turbulent years, and it also characterizes today's situation. Given the increase in the city-wide population, Grünau might face a new turning point. In particular, social stability might be more at risk today because the estate provides cheap housing for low-income households, whereas housing costs in the overall city are currently increasing. The size of flats is also in line with housing policies and allowances for welfare-dependent households—a coincidence of building structures and welfare state policies that might rapidly draw a larger share of lower-status residents.

To conclude, assemblage theory underlines why it is important to understand the pathways of a neighborhood as a sequence of interacting components that might produce contradictions, surprises, and novelties. Assemblage theory is sensitive to taking into consideration external influencing factors, the interplay of factors or dependencies between neighborhoods, and upscale development to explain nonlinearity and uncertainty. The apparently paradoxical development of our case study neighborhoods is a fitting example for the usefulness of this sort of theorizing.

Process and dynamics in context

With the idea of "emergence" comes an emphasis on process and on the dynamics of the social world. In his applications of complexity theory in the field of planning, De

Roo (2010) distinguishes between an analysis of "being" and "becoming," a formulation made famous by Prigogine (1980). This work was transferred to research on neighborhoods and cities, encompassing a shift of attention from the ostensibly static patterns of today's cities to the process of the ongoing production and alteration of these patterns. Here, the (temporary) result cannot be separated from the process. The process, finally, cannot be separated from contexts. Walby conceptualizes "context" as the relationship of a social system to its environment (Walby, 2009, p. 48); for DeLanda (2006), this would be the relationship to other assemblages and their component parts (i.e., external relations). Due to the differences in contexts, the same type of stimulus for development can lead to different outcomes in different places: "Path dependency is summarized aphoristically in the phrase 'history matters' ... [it] holds that a system's trajectory is a function of past states, not just the current state" (O'Sullivan, 2004, p. 285).

Applying this lens to our case studies, we can see that both neighborhoods experienced a change in national political regimes, with Leipzig's inner east experiencing such a change twice. This creates, of course, a huge difference in how properties and characteristics emerge or are further shaped. The emergence of Grünau itself was a top-down planning decision in a state-socialist regime; its subsequent development was situated in a capitalist welfare state. The decisions and practices of housing companies changed and housing became a tradable good. Residents suddenly had new options and new restrictions. Urban planning institutions changed, and so did their assessment criteria for the value of different types of housing stock. Federal urban development programs were formulated on the basis of a combination of western German experiences and the challenges of eastern German cities. In the course of the past 20 years, intermediary and civic actors also entered the arena. The overall population decline in the city (which is the outcome of, again, developments in the wider context) has been a decisive factor for the development of both neighborhoods. Change in either of the areas, for example, is related to residential mobility in the city as a whole. In assemblage wording, the processes in the two neighborhoods are influenced by processes on external developments. Part of the recent influx of new residents is also fueled by the financial crisis, which has made Germany a destination for young unemployed Europeans. The speed and scale of in-migration that Leipzig is experiencing at the moment and the question of whether current population growth will continue or be replaced by new decline will impact the pathways of both Leipzig's inner east and Grünau.

For Leipzig's inner east, the interplay of factors in the wider and local contexts also triggered the neighborhood's movement in seemingly paradoxical directions. The wider context brought investments and support for Leipzig's inner city areas in general. At the local level, *path dependencies* hindered a fast upgrading of the inner east in particular. Leipzig's inner east is a traditional workers' area, with large shares of poor population and a somewhat unfavorable image, which was true even before the Second World War. This explains why the inner east, despite its old built-up stock (which is today largely refurbished) and its excellent location and connectivity, was not among those inner-city districts that saw vibrant upgrading in the late 1990s and 2000s. It also explains why reorganization here has led simply to stabilization and has "only" prevented further decline. However, the high vacancy rates in the refurbished building stock, together with low rents, have made the area very attractive for low-income newcomers: migrants, student flat sharers, young single households, and single-parent households (Haase et al., 2004, 2012). Today's context of Leipzig's housing market explains why Leipzig's inner east will likely become part of one of the next waves of inner-city upgrading—regardless of its negative image and the high concentration of poor households.

To conclude, looking at neighborhood pathways through the lens of assemblage theory makes us aware of the continuous changes in neighborhood patterns and their context-sensitivity. A process orientation shows that any urban moment analyzed by researchers is the result of a (longer) process and should be viewed accordingly. Context-sensitivity in neighborhood research attempts to understand which of many possible development paths exist in a given neighborhood in a certain context, as well as whether one path is more or less probable.

Conclusion

We began with two cases of neighborhood change in post-socialist Leipzig. Our interest was to highlight the complexity and openness of neighborhood processes underway in this setting. In doing so, we hesitated to two of urban theory's major heuristics of neighborhood change: neighborhood decline and gentrification. Instead, we sought to employ complexity and assemblage thinking. Whereas research on neighborhood decline and gentrification is concerned with critically looking at the specific "downward" or "upward" development of a neighborhood, complexity and assemblage thinking are part of an epistemological program that explores oscillating, contradicting, and multidirectional shifts and turns in society.

As stated in the introduction, upon examining our case studies, our concern was that if we look for decline, we are likely to find decline; if we look for gentrification, we are likely to find gentrification. What is more, we observed "downward" and "upward" development in the two neighborhoods occurring simultaneously: for example, rapid out-migration in a period of heavy investments. In a textbook case of decline, upgrading would be a post-migration *reaction*, a renewal strategy. Therefore, we raise the question of how we can produce a more nuanced and comprehensive picture of neighborhood change. With this question, we aim to contribute to recent attempts in urban research to make better sense of contextual differences (see, e.g., Maloutas & Fujita, 2012).

Our conclusion is that complexity and assemblage thinking can—if applied as a first step—facilitate consideration of a fuller picture of the components involved in shaping neighborhood processes and in understanding how these components are interrelated. We suggest, here, an openness to evidence that does not fit the storylines of inherited theories—evidence, for example, that simultaneously supports and contradicts the commonly agreed-upon features of neighborhood decline. With the help of an assemblage approach, we here draw a more relational, context-sensitive, and inductive picture of neighborhood development, with a focus on producing heuristics that explain multidirectional processes and describe the typical relationships and interactions of local components. In this way, we may uncover unexpected dynamics, surprising differences, or countertrends that seem to be stagnant or stable. We can, for example, understand why physical renovation did not lead to sudden upgrading in the case of Leipzig's inner east or why it occurred even in parallel to out-migration in Grünau. In such cases, we have to look at the different components and their interactions not only to understand but also to avoid misinterpreting what we see.

We do not suggest, however, that assemblage theory is the new "one-size-fits-all" approach for analyzing any story of neighborhood change. We are aware that this may lead to a multitude of detailed neighborhood analyses while losing track of common or structural driving forces. In addition, the epistemological frame of complexity and assemblage thinking has yet to be connected to questions of justice, respect, and rights, which we consider one of the strengths of decline and gentrification research (see, e.g., Holm, 2012; Power, 1997). Here, we simply suggest that assemblage approaches may allow researchers to better understand the nuances of neighborhood development that might be otherwise lost. Therefore, it is important to look for anchor points where we can explicitly link the advantages of both epistemologies, being well aware that they operate at different levels and come with different attentions.

Truly bringing together these different approaches is a huge challenge but one that we see as tremendously worthwhile. Part of this challenge is that the questions of power, which are often so decisive for a neighborhood, need to be better integrated into assemblage theory. Among others, Day and Walker (2013, p. 19) attest to a "blindness of assemblage thinking to the structures and uneven power relations within which actants in an assemblage operate." The approach of Allen and Cochrane (2010) on power and policies "of reach" may be a way forward; this approach illustrates the potential of cross-referencing various perspectives on power relations in urban settings, for example, through the lenses of both assemblage theory and gentrification heuristics. The challenges we have described in this paper point to the urgency of new analytical endeavors, as well as theoretical and empirical work, to advance the discourse on neighborhood change.

Disclosure statement

No potential conflict of interest was reported by the author.

References

Allen, John, & Cochrane, Allan (2010). Assemblages of state power: Topological shifts in the organization of government and politics. *Antipode, 42,* 1071–1089.

Anderson, Ben, & McFarlane, Colin (2011). Assemblage and geography. *Area, 43*(2), 124–127.

Beauregard, Robert A. (2012). What theorists do. *Urban Geography, 33*(4), 474–487.

Behling, Michael, & Henn, Sebastian (2010). *Aspekte integrierter Stadtteilentwicklung. Ergebnisse und Erfahrungen aus dem Leipziger Osten.* Berlin: Frank & Timme.

Bernt, Matthias, Haus, Michael, & Robischon, Tobias (Eds.). (2010). *Stadtumbau komplex: Governance, planung, prozess.* Darmstadt: Schader-Stiftung.

Capra, Fritjof (1997). *The web of life: A new scientific understanding of living systems.* New York: Anchor Books.

Cilliers, Paul (1998). *Complexity and postmodernism: Understanding complex systems.* London: Routledge.

City of Leipzig. (2000). *Stadtentwicklungsplan Wohnungsbau und Stadterneuerung.* Leipzig: Author.

City of Leipzig. (2004, 2008, 2010, 2012b). *Ortsteilkatalog 2004, 2008, 2010, 2012.* Leipzig: Author.

City of Leipzig. (2012a). *Monitoringbericht Wohnen 2011.* Leipzig: Author.

City of Leipzig. (2012c). *Statistisches Jahrbuch 2012*. Leipzig: Author.

City of Leipzig. (2014). *Statistischer Quartalsbericht II/2014*. Leipzig: Author.

Clay, Phillip (1979). *Neighborhood renewal*. Lexington, DC: Heath and Company.

Coleman, Alice M. (1985). *Utopia on trial: Vision and reality in planned housing*. Shipman: London.

Davies, Andrew D. (2012). Assemblage and social movements: Tibet Support Groups and the spatialities of political organisation. *Transactions of the Institute of British Geographers, 37*(2), 273–286.

Day, Rosie, & Walker, Gordon (2013). Household energy vulnerability as 'assemblage'. In Karen Bickerstaff, Gordon Walker, & Harriet Bulkeley (Eds.), *Energy justice in a changing climate. Social equity and low-carbon energy* (pp. 14–29). New York: Zedbooks.

De Roo, Gert, Hillier, Jean, & Joris, Van Wezemael (Eds.). (2012). *Complexity and planning: Systems, assemblages and simulations*. Farnham and Burlington: Ashgate.

De Roo, Gert, & Silva, Elisabete A. (Eds.). (2010). *A planner's encounter with complexity*. Farnham: Ashgate.

DeLanda, Manuel (2006). *A new philosophy of society: Assemblage theory and social complexity*. New York: Continuum.

Deleuze, Gilles, & Guattari, Félix (1987). *A thousand plateaus: Capitalism and schizophrenia*. Minneapolis: University of Minnesota Press.

Farías, Ignacio (2010). Introduction: Decentering the object of urban studies. In Ignacio Farías & Thomas Bender (Eds.), *Urban assemblages: How actor-network theory changes urban studies*. New York: Routledge.

Farías, Ignacio, & Bender, Thomas (2010). *Urban assemblages: How actor-network theory changes urban studies*. New York: Routledge.

Gerkens, Karsten, Hochtritt, Petra, & Seufert, Heiner (2010). Städtebauliche Entwicklung. In Michael Behling & Sebastian Henn (Eds.), *Aspekte integrierter Stadtteilentwicklung. Ergebnisse und Erfahrungen aus dem Leipziger Osten* (pp. 71–96). Berlin: Frank & Timme.

Grigsby, William, Baratz, Morton, Galster, George, & Maclennan, Duncan (1987). *The dynamics of neighborhood change and decline*. London: Pergamon.

Haase, Annegret, Kabisch, Sigrun, & Steinführer, Annett (2004). *Results of the questionnaire survey in Leipzig. Re Urban Mobil WP2 Final Report*. Leipzig, 112 pp.

Haase, Annegret, Kabisch, Sigrun, Steinführer, Annett, Bouzarovski, Stefan, Hall, Ray, & Ogden, Philip (2010). Emergent spaces of reurbanisation: Exploring the demographic dimension of inner-city residential change in a European setting. *Population, Space and Place, 16*(5), 443–463.

Hamnett, Chris (2003). Gentrification and the middle-class remaking of inner London, 1961–2001. *Urban Studies, 40*(12), 2401–2426.

Holm, Andrej (2012). Wem gehört die Stadt? Machtkonstellationen in umkämpften Räumen. In Matthias Lemke (Ed.), *Die gerechte Stadt. Politische Gestaltbarkeit verdichteter Räume* (pp. 93–115). Stuttgart: Franz Steiner Verlag.

Holm, Andrej (Ed.). (2014). *Reclaim Berlin: Soziale Kämpfe in der neoliberalen Stadt*. Berlin/Hamburg: Assoziation A.

Ilya, Prigogine (1980). *From being to becoming: Time and complexity in the physical sciences*. San Francisco: W.H. Freeman.

Innes, Judith E., & David. E., Booher (2010). *Planning with complexity: An introduction to collaborative rationality for public policy*. New York: Routledge.

Jessop, Bob (1999). *The governance of complexity and the complexity of governance: Preliminary remarks on some problems and limits of economic guidance*. Retrieved from Lancaster University: http://www.lancs.ac.uk/fass/sociology/papers/jessop-governance-of-complexity.pdf.

Kabisch, Sigrun, & Grossmann, Katrin (2010). *Grünau 2009. Einwohnerbefragung im Rahmen der Intervallstudie „Wohnen und Leben in Leipzig-Grünau".* project report, Leipzig: Helmholtz-Zentrum für Umweltforschung GmbH -UFZ. Retrieved from https://www.ufz.de/export/data/1/49530_publications_grossmann_06_2013.pdf

Kabisch, Sigrun, & Grossmann, Katrin (2013). Challenges for large housing estates in light of population decline and ageing: Results of a long-term survey in East-Germany. *Habitat International, 39*, 232–239.

Ley, David (1994). Gentrification and the politics of the new middle class. *Environment and Planning D: Society and Space, 12*, 53–74.

Maloutas, Thomas, & Fujita, Kuniko (2012). *Residential segregation in comparative perspective: Making sense of contextual diversity.* Farnham, UK: Ashgate.

Manson, Steven M., & O'Sullivan, David (2006). Complexity theory in the study of space and place. *Environment and Planning A, 38*(4), 677–692.

McFarlane, Colin (2011). On context. Assemblage, political economy and structure. *City, 15*(3–4), 375–388.

Musterd, Sako, & Van Kempen, Ronald (2007). Trapped or on the springboard? Housing careers in large housing estates in European cities. *Journal of Urban Affairs, 29*(3), 311–329.

Newman, Oscar (1972). *Defensible space: Crime prevention through urban design.* New York: Macmillan.

O'Sullivan, David (2004). Complexity science and human geography. *Transactions of the Institute of British Geographers, 29*(3), 282–295.

Portugali, Juval, Meyer, Han, & Stolk, Egbert (2012). *Complexity theories of cities have come of age: An overview with implications to urban planning and design.* New York: Springer.

Power, Anne (1997). *Estates on the edge: The social consequences of mass housing in Northern Europe.* London: Macmillan.

Power, Anne, & Tunstall, Rebecca (1995). *Swimming Against the Tide: Polarisation or progress on twenty unpopular council estates 1980–1995.* York: Joseph Rowntree Foundation.

Prak, Niels L., & Priemus, Hugo (1986). A model for the analysis of the decline of postwar housing. *The International Journal of Urban and Regional Research, 10*(1), 1–7.

Rink, Dieter, Haase, Annegret, Grossmann, Katrin, Couch, Chris, & Cocks, Mathew (2012). From long-term shrinkage to re-growth? A comparative study of urban development trajectories of Liverpool and Leipzig. *Built Environment, 38*(2), 162–178.

Skifter-Andersen, Hans (2003). *Urban sores: On the interaction between segregation, urban decay and deprived neighborhoods.* Aldershot: Ashgate.

Smith, Neil (1987). Gentrification and the Rent Gap. *Annals of the Association of American Geographers, 77*(3), 462–465.

Smith, Neil (2002). New globalism, new urbanism: Gentrification as global urban strategy. *Antipode, 34*, 427–450.

Spicker, Paul (1987). Poverty and depressed estates: A critique of "Utopia on Trial". *Housing Studies, 2*(4), 283–292.

STEK LeO – City of Leipzig. (2011). *Integriertes Stadtteilentwicklungskonzept Leipziger Osten.* Leipzig: City of Leipzig.

Thrift, Nigel (1999). The place of complexity. *Theory, Culture & Society, 16*(3), 31–69.

Urry, John (2005). The complexity turn. *Theory, Culture & Society, 22*(5), 1–14.

Van Beckhoven, Ellen, Bolt, Gideon, & van Kempen, Ronald (2009). Theories of neighborhood change and decline: Their significance for post-WWII large housing estates in European Cities. In Rob Rowlands, Sako Musterd, & Ronald van Kempen (Eds.), *Mass housing in Europe: Multiple faces of development, change and response* (pp. 20–50). Basingstoke: Palgrave Macmillan.

Van Wezemael, Joris (2008). The contribution of assemblage theory and minor politics for democratic network governance. *Planning Theory, 7*, 165–185.

Walby, Sylvia (2009). *Globalization and inequalities: Complexity and contested modernities.* Sage: London.

Economic decline and residential segregation: a Swedish study with focus on Malmö

Roger Andersson[a] and Lina Hedman[a,b]

[a]Uppsala University, Institute for Housing and Urban Research, Uppsala, Sweden; [b]Delft University of Technology, Faculty of Architecture and the Built Environment, OTB – Research for the Built Environment, Delft, The Netherlands

ABSTRACT

Economic crises are often associated with increasing levels of income segregation and income polarization. Poor neighborhoods generally hit more severely, with unemployment levels increasing and income levels dropping more than in better-off neighborhoods. In this article, we study the correlation between economic recession and income segregation in Malmö, Sweden, with focus on development in the regions' poorest neighborhoods. We compare and contrast these areas' development during a period of economic crisis (1990–1995) with development during a period characterized by relative economic stability. Our findings suggest that (1) income segregation and income polarization indeed increased during the period of economic crisis; (2) neighborhoods that were already poor before the crisis fared worse than the region in general; and (3) this development was due to both *in situ* changes and to residential sorting, where the differences in income and employment status between people moving into a neighborhood, those moving out, and those who remained in place were greater during the period of recession compared to the more stable period.

Introduction

There is a long scholarly tradition of studying urban development trends, including urban areas' tendencies toward increasing levels of residential segregation and social polarization, in relation to economic change and/or city growth and decline (e.g., Cloutier, 1984; Wilson, 1987). In his overview of ethnic and socio-economic segregation in Europe, Musterd (2005, p. 342) states that "social segregation levels tend to be higher in manufacturing cities and those currently struggling with economic restructuring." However, relatively few studies have attempted to empirically assess how a particular economic transformation – like rapidly expanding unemployment in times of economic crisis – affects residential segregation patterns. It is intuitively expected that economic change will have such effects. During a crisis, households will gain or lose economically and this will translate into spatially selective impacts.

The Swedish economic crisis in 1992–1993 exhibited such geographically uneven outcomes. As shown by Andersson (1998), the Stockholm region lost some 100,000 jobs (about 12% of total employment) from 1990 to 1995; some housing estates, however, lost half of their jobs, and job losses in the range of 30% to 40% were not uncommon in some neighborhoods. The neighborhoods that did comparatively worse were often relatively poor even before the crisis, with an overrepresentation of residents with a low level of education, low earnings, and a vulnerable position in the labor market. Hence, we posit that a financial crisis may lead to increasing levels of segregation and/or increases in the relative concentration of poverty.

A situation in which already poor neighborhoods are becoming even poorer, relative to the city average, can be the result of *in situ* changes in employment rates and income levels (as discussed by Andersson, 1998) or of selective migration processes in which better-off residents leave increasingly poor neighborhoods and are replaced by in-movers with lower socio-economic status. Increasing impoverishment of already poor neighborhoods may also lead to increasing selectivity in mobility patterns, as the poor neighborhoods become even less attractive and stigmatized, which further reinforces the process of decline (Andersson & Bråmå, 2004).

In this article, we ask two sets of questions. The first is related to the relationship between economic decline and residential segregation. Does an economic crisis, such as the Swedish recession during the early 1990s, lead to increasing levels of segregation? Are such patterns more visible in a region hit worse by an economic downturn compared to other regions? We focus on one such region – Malmö – that not only suffered badly from the recession but also, as an older manufacturing region, is struggling with economic restructuring – hence, it should have higher levels of segregation according to Musterd (2005). Our second set of questions examines development in the worst-off neighborhoods in this region and aims to analyze population processes that may enhance the spiral of decline. How did already poor neighborhoods develop during the period of economic crisis? To what extent did their population composition change as a result of the macro-economic situation, including potential increases in residential segregation? To what extent can such changes be explained by *in situ* changes, and what is the role of selective mobility?

Western Europe has recently been hit by a severe recession that has had strong negative effects on the economies of several countries. Sweden has, however, remained relatively unaffected. For example, while unemployment rates in countries like Spain and Greece almost doubled between 2005 and 2010 (2011 in the case of Greece), from about 10% to about 20%, unemployment rates in Sweden remained relatively stable at around 8% (Eurostat). This period of relative stability serves, in this article, as a period of contrast to the financial crisis period of the early 1990s. How did the Malmö region develop in terms of levels of income segregation during this period of stability, compared to the period of crisis? Were there any remarkable differences in the development of the poor neighborhoods in terms of overall socio-economic status, population composition, and residential sorting patterns? To what extent can the patterns identified be attributed to the economic crisis, and to what extent are they part of a more general trend?

Previous research

In a comprehensive study of 216 US metro regions for the period 1970 to 2000, Watson (2009, p. 822) found a "strong and robust relationship between income inequality and income segregation, after controlling for metropolitan area fixed effects, year effects, and a number of other factors. Inequality at the bottom of the distribution is related to the residential isolation of the poor, while inequality at the top is associated with segregation of the rich." She recognized that this relationship, via mechanisms of "neighborhood effects," may also have a reversed causality (i.e., segregation feeds income inequality), but she assumed that "the reverse causality factors are slower-acting and smaller in magnitude than the direct effect of income inequality on residential choice" (pp. 842–843).

Regarding the relationship between economic decline and residential segregation, an important question is whether economic decline leads to increasing inequality. We argue that this is often the case and that it is due to both primary and secondary effects of an economic downturn, at least in the context of an advanced welfare state like Sweden. In short, an economic downturn will have the primary effect of lowering the demand for labor, rendering more people unemployed. This will lower tax revenues and lead to employment cutbacks in the public sector, possibly even to welfare state retrenchment (i.e., less state compensation for unemployed and sick workers). This process describes fairly well the sequence of developments following the acute Swedish financial crisis in the early 1990s, but it should be noted that political decisions did affect developments and that – in theory – a more Keynesian approach to managing the crisis might have resulted in a different chain of events. As in the case of wider European debate today, however, many economists would argue that a Keynesian approach was not possible in the context of a high and rapidly increasing state budget deficit.

Furthermore, researchers have found that financial crises and economic restructuring tend to be harshest for those in the most vulnerable positions in the labor market, while individuals in better positions and/or with higher levels of education generally do better (Clark, 2007; Iceland, Sharpe, & Steinmetz, 2005). Consequently, economic decline often leads to increased economic inequality.

Recent studies on Sweden indicate increasing levels of income inequality (Björklund & Jäntti, 2011; OECD, 2011), and point to the reduced redistributive impact of the tax and welfare benefit systems as the underlying cause of these changes (see also Ferrarini, Nelson, Palme, & Sjöberg, 2012). Following Watson (2009), we anticipate that increasing income inequality in Sweden will lead to an increase in residential segregation by income. Scarpa (2013) considered this complex relationship in a recent study of Malmö, Sweden. Focusing on the 1991–2008 period, he found increasing income inequalities, which he argued were an outcome of the reduced redistributive impact of the Swedish welfare state. He also concluded that increases in residential segregation by income could be attributed to the parallel increase in citywide income inequality rather than to an alleged increase in neighborhood sorting. The latter conclusion is based on the fact that neighborhood income inequality was not associated with a stronger homogeneity of the social composition of neighborhoods in Malmö.

We argue that any attempt to theorize the relationship between economic decline and residential segregation must incorporate theoretical elements from different broad traditions of analysis, including structural as well as behavioral. While structural approaches tend to emphasize the workings of different institutions and markets, they often disregard individual agency. Behavioral approaches focus on households' preferences and decisions and do not explicitly take structural changes into account. However, if, for instance, a poor neighborhood becomes poorer due to rising levels of unemployment, it can be hypothesized that the new conditions will feed secondary changes, such as lowering the level of purchasing power in the neighborhood, which may lead to a reduction of commercial services (Massey & Denton, 1993). Furthermore, deteriorating conditions will also be evaluated by prospective in- and out-movers and will thus influence mobility decisions. This is in line with Wilson (1987, pp. 49–50), who argued that middle-class Black households left the inner city areas of the United States in the wake of worsened employment prospects and social conditions.

As we have noted above, financial crises and economic restructuring tend to be harshest for those in the most vulnerable positions in the labor market (Clark, 2007; Iceland et al., 2005). Consequently, we hypothesize that economic decline will hit already poor neighborhoods, which tend to have an overrepresentation of such inhabitants, especially hard. Many before us have contributed to the field of residential segregation in relation to poverty and poor neighborhoods, focusing on, for instance, out-mobility from socio-economically deprived and/or immigrant-dense neighborhoods and on the effects of neighborhood compositional change (e.g., Meen, Gibb, Goody, McGrath, & Mackinnon, 2005; Van Ham & Clark, 2009; Wilson, 1987). Such patterns and processes are gaining increasing attention in Sweden. For example, Andersson and Bråmå (2004) have shown how deprived neighborhoods are being reproduced through mobility patterns where people who leave these neighborhoods are more often employed than those entering the same areas. This mobility has been found to be more profound during economic downturns than it is during upturns (Andersson, Bråmå, & Hogdal, 2008). Similar patterns have been found in other countries (see Bolt, Van Kempen, & Van Ham, 2008; Card, Mas, & Rothstein, 2008; Jargowsky, 1997).

If the level of segregation increases, it is likely that residents who move will take greater account of neighborhood conditions and that those who have more housing options will tend to avoid particular types of neighborhoods. Such "middle-class flight and avoidance" (Friedrichs, 1998) is generally thought to be based on both "objective" grounds, that is, knowledge about deteriorating social conditions in certain neighborhoods, and on different types of representations (e.g., stigmatization; see Permentier, 2013; Wacquant, 2008). It is furthermore likely that people less aware of a neighborhood's negative associations and/or those who have fewer housing options will become an increasing proportion of residents moving into poor neighborhoods.

As pointed out by Musterd (2005), drawing upon earlier analyses by Galster (1988) in the United States and Peach (1999) in the United Kingdom, ethnic segregation in these countries has socio-economic components, but these sometimes explain only around 10% of the level of ethnic segregation. Swedish studies have indicated a much stronger relationship between ethnic segregation and socio-economic status, in particular a strong overlap between concentrations of poverty and of refugee immigrants (Andersson, Bråmå, & Holmqvist, 2010). Immigrants, especially recent immigrants from refugee countries, tend to have a very weak labor market position and consequently to be highly overrepresented in the poorest neighborhoods. The labor market position and low income among immigrants on average when compared to natives is also commonly cited as one main explanation of ethnic residential segregation in the Swedish context.

All in all, the above discussion suggests that in times of crisis (1) income segregation will increase due to increasing income inequality; (2) immigrants will be more affected than natives, given the former's more vulnerable position in the labor market, which will lead to increased levels of ethnic segregation; (3) this development will on the local (neighborhood) level manifest itself either in more neighborhoods becoming income-poor and immigrant-dense or in these densities increasing in existing poor neighborhoods; and (4) as a behavioral reaction, reinforced selective migration will occur where people (predominantly natives) with higher incomes leave poor neighborhoods and/or avoid moving into such areas.[1]

Data and methods

For this study we make use of rich register data delivered by Statistics Sweden. The GeoSweden database, owned by the Institute for Housing and Urban Research at Uppsala University, comprises all Swedish permanent residents from 1990 to 2010 with annual data on demographic, socio-economic, housing, and work place characteristics. All data are geocoded and all individuals can be followed over time. For the analyses in this article, we use data for 4 years (1990, 1995, 2005, and 2010) split into two five-year periods (1990–1995 and 2005–2010). The first period covers the economic recession of the early 1990s. The second period is used for comparison and represents a period during which much of Europe faced a severe economic crisis while the Swedish economy remained relatively stable. Changes that took place in the earlier period but not in the later one can hence be assumed to be at least partly due to the 1990s recession.[2] Since our focus is on labor market–related outcomes, we have restricted the population to individuals of working age, 20–64 years.

Our study employs two different spatial units, the labor market region and the neighborhood. Sweden is divided into 100 labor market regions, which are administrative areas based on commuting patterns. They vary substantially in size and population. Malmö labor market region encompasses 15 municipalities and was in 2010 home to about 700,000 people. On the labor market region, we present descriptive data for income segregation levels in 1990 and 2010, complemented with more detailed information about population composition, employment levels, income polarization (understood as population shares belonging to the lowest, middle, and highest income groups, defined below) and levels of segregation for Stockholm and Malmö in 1990, 1995, 2005, and 2010. Estimates of segregation and polarization are based on differences between neighborhoods within the region. Neighborhoods are defined as SAMS (small area market statistics) units, having on average around 1000 residents. The SAMS area division is made by Statistics Sweden in collaboration with each municipality. SAMS is a frequently used proxy for neighborhoods in Swedish segregation and neighborhood effect studies (e.g., Bråmå, 2006).

Our main focus, however, is not on neighborhoods or segregation patterns in general but on (population) processes leading to increasing levels of poverty concentration in the region as a whole and in the worst-off neighborhoods in particular. Most of our analyses focus on such processes in Malmö's poorest neighborhoods. We have selected neighborhoods that were targeted by the political interventions launched in the mid-1990s to avoid and even thwart further segregation through a range of efforts aimed at improving liveability, employment, school children's academic performances, and cultural life (Andersson, 2006).

Key variables in this study are income, employment status, and immigrant status. Income is defined as income from work, including income from work-related benefits such as parental leave, sick leave, and so forth. In a welfare state of the pre-crisis Swedish type the link between work income and poverty, defined according to disposable income, was relatively weak.[3] Despite this, we argue that income from work is a better estimate of a person's socio-economic status since it not only measures income but also, indirectly, employment status, type of job, and level of education; furthermore, it is unaffected by state transferences such as child benefits, housing allowances, and social benefits. The correlation between economic decline/development and income should consequently be more evident by using income from work. Based on the national income distribution, we categorize individuals into three groups (low, middle, and high income) according to income decile: low = decile 1–3, middle = decile 4–7, high = decile 8–10. A similar categorization has been used in several other studies analyzing income segregation and its consequences (see Galster, Andersson, & Musterd, 2014).

Employment is measured yearly in the first week in November. In order to exclude persons with only temporary contracts, we define someone as employed if he or she is registered as employed and has an annual earning above the amount computed by Statistics Sweden for social insurance purposes ("basbelopp") – SEK 29,700 in 1990 and SEK 42,400 in 2010.

For each person in the data, we also have information on country of birth. For simplicity, we categorize country of birth into four large groups: Swedish (including Swedish-born children of immigrants), Western (individuals born in countries of the EU15 and EES, and in North America, Japan, and Oceania), Eastern European (born in the rest of Europe and in Russia and former Soviet states), and non-Western (born in the rest of Asia including Turkey, and in Africa and Latin America). A majority of refugee countries are found in the latter category; the main exception is the large inflow of people from former Yugoslavia during the mid-1990s.

Most of our analyses are descriptive in character and are presented in tables and figures below. Descriptive information provides an overview of the trends during the two periods. However, to get a better understanding of the residential sorting in Malmö's poor neighborhoods, we complement the descriptive analyses with two logistic multivariate regression analyses. The objective of these regressions is to identify key characteristics of groups we define as "in-movers," "stayers," and "out-movers" in respect to poor Malmö neighborhoods. The regression analyses include the above-described variables together with control variables related to individuals' sex, age, family status, and level of education. In the first model, analyzing differences between in-movers to the poor areas and in-movers to other areas, we include a dummy variable indicating whether the individual moved within or into the Malmö region. In the second analysis, analyzing differences between stayers and out-movers from the poor areas, all individuals lived in the region by default. Unlike the first regression, however, this second regression includes variables that indicate changes in family status and income during the five-year period. Income change is categorized into quartiles (quartile 1 is the 25% of our population having the weakest income development 1990–1995/ 2005–2010 and quartile 4 the 25% seeing the most improvements). Change in family composition is estimated as going from single to couple and vice versa, and from not having children living in the household to having children and vice versa. Such changes in family composition and income are generally important in explaining mobility decisions. These variables could not, however, be included in the first regression due to a lack of data for some individuals – not all in-movers to Malmö's neighborhoods lived in Sweden in the first year of the period (1990 or 2005), and since we argue that the inflow of new immigrants is key to understanding the reproduction of the characteristics of these worst-off neighborhoods, it would be unwise to exclude them. For the same reason – the many newcomers for whom we do not have information for the first year of the period – we estimate many control variables at $t + 5$, that is, at the end of the period. The reader should thus be aware that the findings do not necessarily represent the status of the individuals at the time they moved in/out.

Employment and segregation change in Sweden 1990 to 2010

Due to combinations of political economic reforms (financial deregulations), a bank crisis, currency policies, and international market changes, Sweden faced a severe economic downturn in 1992 to 1993/1994. GDP fell by 5.1% between 1991 and 1993, and private investments by 35% (Englund, 1999). The recession eventually affected all economic sectors and led to severe employment cutbacks, first in the export sectors and later in sectors relying on domestic demand and tax revenues (Giavazzi & Pagano, 1996); registered unemployment increased from about 1.5% to 8.2% (Ekonomifakta, 2013; see also Calmfors, 1994, p. 7). Unemployment rates have been fairly stable at that level since, and it seems highly unlikely that the pre-crisis employment levels will be re-established in the foreseeable future.

Regional variations were quite marked. Employment levels were relatively low in parts of the north and the southeast of the country and very high in a number of towns dominated by manufacturing industry and in the large metropolitan regions of Stockholm, Malmö, and Gothenburg. Among the larger cities, Malmö had without doubt the weakest economic development (it ranked third from the bottom among the 100 labor market regions 1990 to 2010). Since 1990, Malmö has had a very limited increase in job opportunities (from 278,000 to 295,000), while the population increase among working-age people has been substantial. This has resulted in a sharp overall reduction in employment frequencies, from 83% employed in 1990 to below 73% 20 years later.

Segregation by income is also generally higher in the larger metropolitan regions. Although income segregation is generally low in Sweden due to a combination of housing and urban policies, wage policies, and redistribution of welfare through the tax and benefit systems, a vast majority of the labor market regions experienced an increase in income segregation between 1990 and 2010. The average index of segregation measured as the residential differences between the three lowest work income

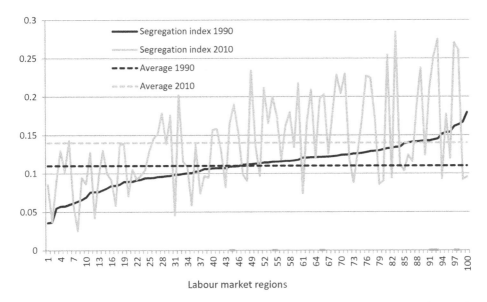

Figure 1. Index of segregation (low vs. middle and high income) in 1990 and 2010 for Sweden's labor market regions.

deciles and the remaining seven deciles stood at .11 in 1990 and increased to .14 twenty years later (see Figure 1). Regions hit hard by the economic crisis, resulting in an increase in the share of low-income people, also experienced an increase in the level of income segregation. A similar pattern was not found after the crisis (1995–2010).

Table 1 provides more detailed information on trends in population composition, employment, income polarization, and segregation for Malmö and Stockholm (the latter for comparative purposes). It reveals that the severe drop in employment rates took place during the economic crisis (1990–1995), whereas employment levels have been relatively stable ever since. Income polarization, as measured by the Gini index, increased during the crisis in both Stockholm and Malmö, but has remained relatively stable since the crisis (see also Andersson et al., 2010). Income segregation has developed similarly in the Malmö case, starting at a low initial level, increasing during the crisis years, and then remaining relatively stable at the new higher level. Both Stockholm and Malmö have witnessed a profound change in ethnic population composition between 1990 and 2010. In Malmö, the share of foreign-born people in the working-age population increased from 14% to almost 26% and the share of people born in non-Western countries grew from 3% to 11%. This trend mirrors a general trend in Sweden, with a very fast rate of refugee immigration. Such a rapid change in ethnic composition of the population might be relatively unproblematic in periods of prosperous economic development, but in times when unemployment levels are high it is generally very difficult for the new immigrants to find jobs. As shown in Table 1, employment frequencies are substantially lower among immigrants in general and non-Western immigrants in particular in both Stockholm and Malmö. The non-Western immigrants also suffered more from the financial crisis, with larger drops in employment rates. They have, however, also recovered better than the population in general.

Table 1. Population, employment, polarization, and segregation change in Stockholm and Malmö 1990 to 2010.

		Stockholm				Malmö			
Aspect	Variable	1990	1995	2005	2010	1990	1995	2005	2010
Population	Total population (1000s)	1,960	2,067	2,245	2,417	584	615	661	716
	Population aged 20 to 64	1,184	1,256	1,377	1,465	342	362	398	430
	% aged 20 to 64	60,4	60,8	61,4	60,6	58,5	58,9	60,2	60,0
	Thereof foreign-born (20 to 64)	218	246	301	372	49	60	84	110
	% foreign-born (20 to 64)	18,4	19,6	21,9	25,4	14,3	16,6	21,0	25,6
	Thereof non-West (20 to 64)	66	95	157	211	11	18	31	46
	% non-West of total (20 to 64)	5,6	7,6	11,4	14,4	3,3	4,8	7,8	10,8
Employment	Percent employed (20 to 64)	84,5	72,9	75,3	76,5	81,4	67,6	69,2	68,6
	Percent foreign-born employed (20 to 64)	72,7	52,6	58,0	59,8	64,2	40,4	45,3	44,4
	Percent non-West employed (20 to 64)	63,9	39,6	52,6	56,0	51,7	27,7	37,3	37,7
Social	Bottom 3 work income deciles share of total income	6,1	2,5	3,0	2,5	8,8	3,2	3,6	3,1
polarization	Middle 4 work income deciles share of total income	32,4	31,7	28,0	29,5	39,7	39,7	38,7	36,3
	Top 3 work income deciles share of total income	61,5	65,9	69,1	68,0	51,5	57,1	57,7	60,6
	Gini coefficient work income distribution	0,18	0,23	0,23	0,23	0,19	0,24	0,24	0,25
Segregation	Work income segregation (IS, low vs. rest (20 to 64))	0,14	0,18	0,17	0,21	0,16	0,25	0,25	0,26
	Ethnic segregation (ID, non-West vs. Sweden (20 to 64))	0,47	0,51	0,49	0,48	0,53	0,55	0,52	0,51
Neighborhoods	Number of neighborhoods (SAMS)	1,336	1,352	1,271	1,265	778	789	771	778
	Average number of residents per SAMS	886	876	1,152	1,157	750	779	858	920

IS = index of segregation; ID = index of dissimilarity.

Zooming in: income segregation, inequality, and residential sorting in Malmö and its poorest neighborhoods

Malmö is Sweden's third largest labor market region, home to approximately 716,000 inhabitants in 2010. The region has long been dominated by the manufacturing industry. The decline in such jobs over the past 30-plus years has thus had a great impact on the region's economic development; as we have shown, unemployment levels are generally higher in Malmö compared to other large Swedish cities and the region was also more negatively affected by the economic recession in the early 1990s. Like many other European formerly industrial cities, the Malmö region has also witnessed a shift in its population distribution. Over time, Malmö city has experienced a relative population decline, while suburban municipalities in the region have attracted many households that have managed relatively well during the deindustrialization phase. This shift can partly be explained by the tenure structure of the region, where the core is dominated by multifamily housing and home ownership dominates the suburbs. The consequence of this shift is that Malmö city has been more negatively affected by the sluggish economic development than have its surrounding suburbs.

The concentration of low-income people in the regional core is one explanation for the Malmö region's increasing levels of income segregation. Another is the change within the core towards increasing concentrations of low-income people in certain neighborhoods. Between 1990 and 2010 the region saw a sharp increase in the number of people living in a low-income neighborhood, defined as a neighborhood where the share of low-income people is two standard deviations higher than the neighborhood average of the region. This increase was due to both a general increase in the number of neighborhoods defined as poor and an increase in the number of inhabitants in the already poor neighborhoods. Between 1990 and 1995, the number of poor neighborhoods, as defined above (and only including neighborhoods with a minimum of 30 inhabitants), increased from 16 to 24, or from 2.4% to 3.6% of the total number of neighborhoods in the region. At the same time, the number of inhabitants living in poor neighborhoods increased from 2.7% of the region's population in 1990 to 6.7% in 1995. Most of these were low-income people.

A trend towards increasing levels of segregation and/or increasing spatial concentration of a subgroup can be explained by two different sets of population dynamics. The first is a change in existing neighborhood populations (for a discussion of the importance of such processes related to ethnic segregation, see Finney & Simpson, 2009). Inhabitants living in already poor neighborhoods, or in neighborhoods close to poverty, may for instance be more severely affected by an economic recession than inhabitants in other neighborhoods. The second type of explanation relates to selective migration, where employment levels and/or income levels among in-movers to poor neighborhoods are well below those of out-movers.

In the subsequent sections, we will analyze population dynamics in the poorest neighborhoods in more detail during the two periods. For this section we have selected neighborhoods that were targeted by the political interventions launched in the mid-1990s. We do not aim to evaluate the efforts as such but can conclude that the outcome has been very different from the hopes stated in official documents. Rather than improving the economic situation of inhabitants, development in these areas until 2010 has been characterized by a substantial increase in the proportion of people with very low incomes, while the presence of better-paid workers has declined (see Table 2). Employment levels have also decreased dramatically, from 71.6% in 1990 to 44.5% in 2010. However, in line with the general economic trajectory, this downward trend occurred mainly during the economic crisis in the early 1990s. Between 1990 and 1995, employment levels in these neighborhoods fell by an astonishing 30 percentage points, to almost half the original level, and the share of low-income people increased from 43% to 60.3%. Since 1995, these proportions have been relatively stable. The

Table 2. Population, employment, and income polarization in Malmö's targeted neighborhoods, 1990, 1995, 2005, and 2010.

| Aspect | Variable | Targeted neighborhoods | | | |
		1990	1995	2005	2010
Population	Total population	61,430	64,029	73,080	78,774
	Population aged 20 to 64	34,012	37,033	43,832	49,205
	% aged 20 to 64	55.4	57.8	60.0	62.5
	Thereof foreign-born (20 to 64)	12,464	17,104	24,911	30,413
	% foreign-born (20 to 64)	36.6	46.5	56.8	62.0
	Thereof non-West (20 to 64)	4,134	6,981	12,823	17,221
	% non-West of total (20 to 64)	12.2	18.9	29.3	35.0
Employment	% employed (20 to 64)	71.6	41.4	44.7	44.5
	% foreign-born employed (20 to 64)	57.1	23.2	33.8	34.5
	% non-West employed (20 to 64)	47.3	16.6	26.8	28.3
Social polarization	Bottom 3 work income deciles share of total income	43.0	60.3	61.5	61.9
	Middle 4 work income deciles share of total income	41.2	28.6	29.9	30.1
	Top 3 work income deciles share of total income	47.3	11.1	8.6	8.0
Neighborhoods	Number of neighborhoods	49	49	49	49
	Average number of residents per neighborhood	1,307	1,306	1,491	1,608

number of people living in the targeted areas has, however, increased substantially over time.

Another clear trend is the change in ethnic composition. In 1990 immigrants from non-Western countries made up 12.2% of the working age population in the targeted neighborhoods. Between 1990 and 1995, this share went up by an astonishing 55%, accompanied by an increasing geographical concentration of the immigrant population in the core city and especially in the poor neighborhoods (see also Andersson et al., 2008; Billing, 2000). The share of non-Western immigrants living in the targeted areas has continued to increase over time, and in 2010 35% of the working-age population in these neighborhoods was born in a country classified as non-Western. In the same year 62% of the population was born abroad. One explanation is the construction of the Öresund Bridge between Sweden and Denmark, which has resulted in an increasing Danish-born population, but more important in this context is that Malmö is a popular destination for newly arrived refugees.

The poor development during especially the early time period can, as mentioned, be explained either by the fact that the existing population in the targeted areas was hit harder by the crisis than inhabitants in other neighborhoods, or by selective migration into and out of these areas. In Figure 2 we show employment status in 1990 and 1995 and in 2005 and 2010 for individuals who lived in the same neighborhood in both years, for targeted areas and the entire Malmö labor market region (targeted areas included). The figures show clearly that employment rates are considerably lower in the targeted areas compared to the region and that this gap is larger in the later period. However, while there is not much difference between the targeted areas and the region in terms of changes in employment during the more recent period, Figure 2 shows that inhabitants in the targeted areas were indeed more severely affected by the economic recession in the early 1990s. More people living in these areas went from employment to none-mployment compared to those living in other parts of the region (21.2% vs. 13.7%). This development enhanced the initial gap in employment rates by 14 percentage points. However, although over two-fifths of the employed "stayers" in targeted neigh-borhoods lost their jobs between 1990 and 1995, the overall employment rate in these neighborhoods fell even more (as shown in Table 3). Hence, it is plausible that selective migration patterns further reinforced the downward trend in these areas.

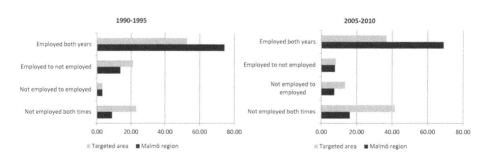

Figure 2. Employment rates in 1990–1995 and 2005–2010 for stayers in targeted neighborhoods and all of Malmö labor market region for population aged 20–60 in 1990 (25–65 in 1995). Values in %.

Table 3. Descriptive statistics for variables used in the regression analyses (mean values). All variables are dummy variables with min = 0, max = 1.

	Moving into targeted neighborhoods		Moving into other neighborhoods		Staying in targeted neighborhoods		Moving out of targeted neighborhoods	
	1990–1995	2005–2010	1990–1995	2005–2010	1990–1995	2005–2010	1990–1995	2005–2010
Sex (male = 1)	0.54	0.55	0.52	0.50	0.49	0.49	0.51	0.52
Age 34+	0.40	0.29	0.39	0.43	0.64	0.63	0.42	0.36
New in Malmö	0.20	0.13	0.16	0.16				
Single with child(ren)	0.10	0.07	0.10	0.09	0.14	0.14	0.10	0.09
Couple without child(ren)	0.43	0.11	0.42	0.10	0.38	0.10	0.36	0.08
Couple with child(ren)	0.10	0.28	0.11	0.39	0.15	0.39	0.09	0.43
Change in family status					0.25	0.25	0.35	0.43
Born in Western country	0.06	0.09	0.05	0.04	0.07	0.04	0.08	0.04
Born in Eastern European country	0.28	0.16	0.05	0.06	0.19	0.25	0.14	0.17
Born in non-Western country	0.27	0.39	0.04	0.06	0.14	0.33	0.10	0.22
Employed	0.30	0.40	0.70	0.79	0.52	0.50	0.55	0.68
Low income	0.72	0.66	0.31	0.25	0.49	0.54	0.35	0.38
Middle income	0.20	0.26	0.40	0.40	0.36	0.35	0.36	0.43
Income change quartile 1					0.32	0.18	0.29	0.21
Income change quartile 2					0.22	0.47	0.13	0.28
Income change quartile 3					0.29	0.17	0.24	0.17
Low education	0.39	0.27	0.20	0.17	0.43	0.44	0.26	0.27
Middle education	0.43	0.34	0.54	0.46	0.47	0.39	0.49	0.42
Valid N (list-wise)	13,142	16,319	104,448	107,200	18,559	25,163	15,453	13,330

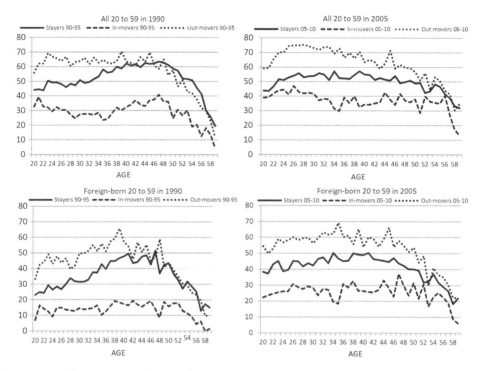

Figure 3. Employment rates by age for targeted Malmö neighborhoods in 1995 and 2010, by stayers, out-movers, and in-movers with respect to these areas in 1990 to 1995 and 2005 to 2010. Top two: All in working ages. Bottom two: foreign-born in working ages.

Figure 3 shows employment levels among stayers, in-movers, and out-movers with respect to the targeted areas during our two time periods. The general pattern is that those moving into the targeted areas had a substantially lower employment rate than those leaving and – especially among younger people – also lower than those staying, which is in line with previous Swedish findings (Andersson & Bråmå, 2004). This applies to immigrants as well as natives. The gap between out-movers and in-movers was, however, larger during the earlier period, mostly due to a lower employment rate among in-movers in 1990–1995 compared to 2005–2010. A potential interpretation is that people who became unemployed during the crisis, or whose employment prospects decreased during this period, became more likely to move to these neighborhoods. Such patterns, combined with outflows of employed individuals, would lead to increased concentrations of poverty over time. Before drawing such conclusions, however, it is important to look at where the in-movers came from – whether from within the region, or from other parts of Sweden, or abroad. Another difference between the two periods concerns employment rates by age. It seems as if the more recent period posed a much more difficult employment situation for older people. While employment rates in 1995 peaked at around age 50, they now peak much earlier (35 to 40 for stayers and around 25 for the entire group of out-movers).

In an attempt to cast more light on the sorting going on in Stockholm's immigrant-dense neighborhoods, Andersson (2013) studied all intra-urban migrants and neighborhood stayers 2005–2008. He found ethnic differences in the probability of entering

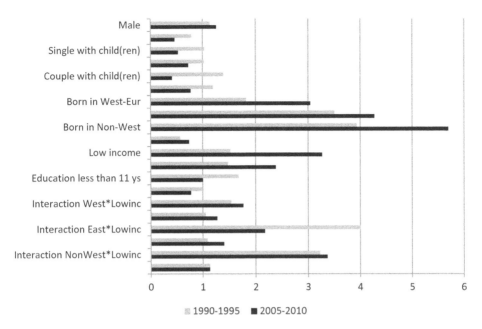

Figure 4. Odds ratios from regression comparing movers into targeted neighborhoods (1) to movers elsewhere in the region (0). Most values are statistically significant, with the exception of those very close to 1 (see Appendix Table A1).

immigrant-dense neighborhoods but not for the probability of leaving, which he argued supports the "native avoidance" but not the "native flight" hypothesis (see also Bråmå, 2006). We follow up on this analysis by conducting similar analyses for the targeted Malmö neighborhoods, but with more focus on sorting based on income and employment. We run two different models. The first analyses differences between in-movers to the targeted neighborhoods and in-movers to other neighborhoods in the Malmö region and the second analyses differences between stayers and out-movers with respect to the targeted areas themselves. Both models are run twice, once for each time period. Descriptive statistics of all variables are found in Table 3. Results of the two models, for the two periods, are shown in Figures 4 (model 1, comparing in-movers to the targeted area to "other in-movers") and 5 (model 2, comparing stayers to out-movers). Full results from the statistical models, including standard errors and level of significance, are in the Appendix.

Odds ratios from the model comparing movers into targeted neighborhoods to movers moving elsewhere in the Malmö region are presented in Figure 4. Results show that movers into the targeted areas were less likely to be employed and considerably more likely to have a low income compared to other movers. They were also considerably more likely to be born outside of Sweden. In addition, results show that these differences are generally more pronounced in the later period, with the exception of differences in employment. The poor economic development during the earlier period thus resulted in increased sorting based on employment, while other aspects of sorting – especially income and ethnicity – became more important later. In addition, the combination of low income and ethnic minority status results in substantially

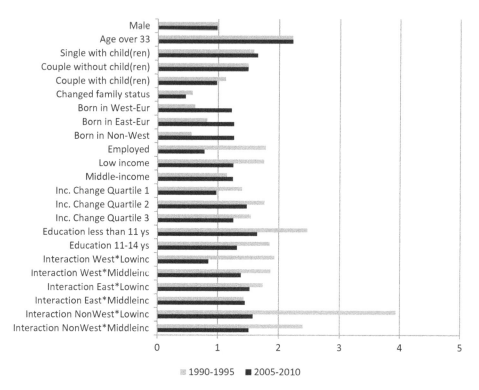

Figure 5. Odds ratios from regression comparing stayers in targeted neighborhoods (1) to out-movers from targeted neighborhoods (0). Most values are statistically significant, with the exception of those very close to 1 (see Appendix Table A2).

higher odds of entering the targeted areas. In 1990–1995 this effect is strongest for immigrants from Eastern Europe, probably explained by the large refugee immigration from the former Yugoslavia, while it is stronger for people from non-Western countries during the later period.

Figure 5 presents odds ratios from the regression comparing stayers in the targeted neighborhood to out-movers from the same areas. Since this model only includes people who lived in the targeted areas (and hence in Sweden) in the first year (1990 or 2005), we also include some estimates of change in family status and income over time. In the early period, we see that stayers in the targeted neighborhoods were more likely than out-movers to be employed at the end of the period but also more likely to have a low level of education and a low level of income and to have experienced a drop in income (income change quartile 1 or 2) between 1990 and 1995. The odds ratios for being born abroad are all negative, but the interaction terms reveal that this is only relevant for those with high incomes – all ethnic minority groups are more likely than Swedish born to remain in the targeted areas if their incomes are at a low or medium level, and this is especially true for people of non-Western origin. The socio-economic patterns also hold true during the second period, 2005–2010, but they are generally less strong. Hence, it appears that the economic recession also had a strong impact on sorting patterns from the targeted areas, leaving those with low and/or dropping incomes behind. The results for ethnic minority groups of low or medium incomes are, however, very similar compared to the early time period (ethnicity term and interaction term combined; the exception is non-Western minorities with low income, for whom the odds of staying put were much stronger during the first period). Ethnic minorities below the high income level thus tend to be "trapped" in the targeted areas to a much higher extent than low-income natives and/or high-income minorities.

Overall, we find that sorting on economic grounds is stronger during the economic recession compared to the more recent period, when people of low incomes were more likely to both move into and remain in our case study areas. The extra penalty for low-income ethnic minorities exists during both time periods but we clearly see how the inflow of new migrants affects sorting patterns – during the early period, low income Eastern Europeans were more likely to move into the targeted areas, while non-Western immigrants were more likely to do so during the second period. This is likely due to a shift in immigration patterns, where refugee immigration went from being dominated by former Yugoslavia during the early period to countries like Iraq, Syria, and Somalia during the later period. Ethnic differences are generally more pronounced for in-migration patterns into the targeted areas than for out-migration patterns from the same areas, especially during the second period, which is in line with previous findings (Andersson, 2013; Bråmå, 2006).

Discussion and conclusions

In this article, we study the link between economic recession and levels of income segregation. Hypothesizing that those in the most vulnerable position on the labor market are hit hardest by an economic crisis, and therefore, that the consequences of a crisis are also more serious in neighborhoods that were already poor before the recession, we zoom in on a number of such neighborhoods in the Malmö region and analyze population processes that may enhance such a trajectory of decline. We do this by looking at two different time periods – one that was characterized by a severe economic crisis and one during which the economy was relatively stable. Although we cannot, based on our data, establish that economic recession *causes* increased levels of segregation and concentrations of poverty, our results suggest that there is a clear correlation between these two. In short, the period of economic recession in the early 1990s was characterized by sharp increases in both income inequality and income segregation and this development was especially pronounced in regions that suffered badly from the recession. The already poor neighborhoods experienced more dramatic increases in unemployment rates and relative share of low-income people compared to neighborhoods that were better off when the crisis began. During the more economically stable period, segregation levels are also more stable. These conclusions are in line with previous research, for example, results by Musterd (2005) stating that regions struggling with economic restructuring tend to have higher levels of residential segregation, and by scholars finding correlations between increasing levels of income inequality and increasing levels of income segregation (Scarpa, 2013; Watson, 2009). It should, however, be added that this is certainly not a necessary outcome. The effects depend on a range of contextual issues, most importantly how the level of income redistribution (i.e., welfare) nationally and locally are affected by reduced levels of work income and higher rates of unemployment, as well as key features of the local housing markets (e.g., how housing prices change and how neighborhoods are composed in terms of tenure, dwelling sizes, and housing types). Sweden in 1990 had a compressed social system and a perhaps even more compressed system of neighborhoods (Musterd & Andersson, 2005). Many neighborhoods were socially mixed and the level of income segregation was modest. Despite subsequent negative developments in terms of income segregation, levels of segregation were still modest in 2010 compared to many other countries.

Zooming in on the development in the poorest neighborhoods in the Malmö region, we found increasing numbers and proportions of low-income residents and found that the number of poor neighborhoods increased. We furthermore concluded that the share of unemployed and low-income people increased in the "old" poor neighborhoods. This negative trend in the poorest segment of the housing market is due to a combination of inhabitants in these areas being affected especially severely by the crisis – they were, for example, more likely than others to lose their jobs – and residential sorting. The status of the poor neighborhoods is constantly reproduced through negative socio-economic selection in which migrants to these neighborhoods are much poorer than other migrants to and within Malmö and those leaving these areas have a stronger position in the labor market than those entering and staying. The economic aspects of sorting were much more pronounced during the period of crisis compared to the period of economic stability, suggesting a correlation between macro-level economic development and rising levels of poverty in the poorest areas.

This sorting also has a distinct ethnic component. In Sweden, and in the poorest segments of the housing market especially, it is difficult to disentangle effects of economic segregation and ethnic segregation, since the poor areas tend to have a clear overrepresentation of ethnic minorities who often have a very weak position in the labor market. However, our findings suggest that the ethnic sorting was fairly similar during both periods. Low-income immigrant groups were generally more likely than both low-income natives and high-income immigrants to both enter the poorest neighborhoods and to remain there. These findings are in line with those of Andersson (2013) and Bråmå (2006) and suggest that these poor, immigrant-dense neighborhoods are being reproduced over time through the in-migration of newly arrived immigrants and a "flight" of comparatively better-off individuals.

In sum, our results suggest that the economic crisis was indeed associated with rising levels of income segregation and income inequality and that regions and neighborhoods that were already doing relatively badly before the crisis tend to be hit harder than other areas. The crisis took place at a time when the Swedish welfare state was stronger than it is today, with more effective welfare transfers and redistributions and thus a weaker link between work income and poverty. Initial levels of income inequality were also considerably lower. The structural changes that have taken place since the 1990s suggest that an economic crisis today would have an even stronger effect on income inequality and income segregation, resulting in higher levels of segregation and a further enhancing decline in the already worst-off neighborhoods.

Acknowledgment

Roger Andersson is grateful for the generous support provided for the academic year 2013–2014 by New York University School of Law.

Disclosure statement

No potential conflict of interest was reported by the authors.

Funding

This research has received funding from the European Research Council under the European Union's Seventh Framework Program [FP/2007-2013]/ERC Grant Agreement n. 615159 (ERC Consolidator Grant DEPRIVEDHOODS, Socio-spatial inequality, deprived neighborhoods, and neighborhood effects).

Notes

1. The applicability of the White Flight hypothesis to the Swedish case has, however been questioned by Andersson (2013) and Bråmå (2006).
2. It is, of course, possible that other structural changes have had an effect on out outcomes; therefore, we can only make inferences about this economic development and cannot draw any firm conclusions about causality.
3. However, from 1990 onward, many political decisions have made the link between work and disposable income more direct. For example, unemployment now translates more rapidly into reduced disposable income as unemployment benefits are now less generous and last a shorter period of time. For Malmö, this becomes evident when we compare the relationship between the proportion of lowest work and disposable income quintiles in poor neighborhoods in 1990 and 2010: in 1990, 35% of people aged 20 to 64 were in the lowest work income quintile while only 18% were in the lowest disposable income quintile; for 2010, the corresponding proportions were 53% and 50%, respectively.

References

Andersson, Roger (1998). Segregering, segmentering och socio-ekonomisk polarisering. Stockholmsregionen och sysselsättningskrisen 1990–95. Partnerskap för Multietnisk Integration, Rapport nr. 2, Sociologiska institutionen, Umeå universitet.

Andersson, Roger (2006). 'Breaking segregation' – Rhetorical construct or effective policy? The case of the metropolitan development initiative in Sweden. *Urban Studies, 43*(4), 787–799.

Andersson, Roger (2013). Reproducing and reshaping ethnic residential segregation in Stockholm: The role of selective migration moves. *Geografiska Annaler: Series B, Human Geography, 95*(2), 163–187.

Andersson, Roger, & Bråmå, Åsa (2004). Selective migration in Swedish distressed neighbour-hoods: Can area-based urban policies counteract segregation processes? *Housing Studies, 19* (4), 517–539.

Andersson, Roger, Bråmå, Åsa, & Hogdal, Jon (2008). *Fattiga och rika – segregerad stad. Flyttningar och segregationens dynamik i Göteborg 1990–2006*. Gothenburg: Stadskansliet.

Andersson, Roger, Bråmå, Åsa, & Holmqvist, Emma (2010). Counteracting segregation: Swedish policies and experiences. *Housing Studies, 25*(2), 237–256.

Billing, Peter. (2000). *Skilda världar? Malmös 1990-tal i ett kort historiskt perspektiv*. Malmö: Malmö City.

Björklund, Anders, & Jäntti, Markus (2011). *Inkomstfördelningen i Sverige*. SNS Välfärdsrapport 2011. Stockholm: SNS Förlag.

Bolt, Gideon, Van Kempen, Ronald, & Van Ham, Maarten (2008). Minority ethnic groups in the Dutch housing market: Spatial segregation, relocation dynamics and housing policy. *Urban Studies, 45*(7), 1359–1384.

Bråmå, Åsa (2006). "White Flight"? The production and reproduction of immigrant concentration areas in Swedish cities, 1990–2000. *Urban Studies, 43,* 1127–1146.

Calmfors, Lars (1994). What can Sweden learn from the European unemployment experience? An introduction. *Swedish Economic Policy Review, 1,* 5–26.

Card, David, Mas, Alexandre, & Rothstein, Jesse (2008). Tipping and the dynamics of segregation. *Quarterly Journal of Economics, 123*(1), 177–218.

Clark, W. A.V. (2007). Race, class, and place: Evaluating mobility outcomes for African Americans. *Urban Affairs Review, 42*(3), 295–314.

Cloutier, N R. (1984). The effect of structural and demographic change on urban residential segregation. *Review of Social Economy, 42*(1), 32–43.

Ekonomifakta (2013). Structural problems and reforms. Retrieved September 24, 2013, from http://www.ekonomifakta.se/en/Swedish-economic-history/Structural-Problems-and-Reforms/

Englund, Peter (1999). The Swedish banking crisis: Roots and consequences. *Oxford Review of Economic Policy, 15*(3), 80–97.

Ferrarini, Tommy, Nelson, Kenneth, Palme, Joakim, & Sjöberg, Ola (2012). *Sveriges socialförsäkringar i jämförande perspektiv. En institutionell analys av sjuk-, arbetsskade- och arbetslöshetsförsäkringarna i 18 OECD-länder 1930 till 2010.* Stockholm: Socialdepartementet.

Finney, Nissa, & Simpson, Ludi (2009). *"Sleepwalking to segregation"? Challenging myths about race and migration.* Bristol: Policy Press.

Friedrichs, Jürgen (1998). Do poor neighbourhoods make their residents poorer? Context effects of poverty neighbourhoods on their residents. In H. Andress (Eds.), *Empirical poverty research in a comparative perspective* (pp. 77–99). Aldershot: Ashgate.

Galster, George (1988). Residential segregation in American cities: A contrary review. *Population Research and Policy Review, 7,* 93–112.

Galster, George, Andersson, Roger, & Musterd, Sako (2014). Are males' incomes influenced by the income mix of their male neighbors? Explorations into nonlinear and threshold effects in Stockholm. *Housing Studies, 30*(2), 315–343.

Giavazzi, Francesco, & Pagano, Marco (1996). Non-Keynesian effects of fiscal policy changes: International evidence and the Swedish experience. *Swedish Economic Policy Review, 3*(1), 67–103.

Iceland, John, Sharpe, Cicely, & Steinmetz, Erika (2005). Class differences in African American residential patterns in U.S. metropolitan areas: 1990–2000. *Social Science Research, 34*(1), 252–266.

Jargowsky, Paul A. (1997). *Poverty and place: Ghettos, barrios and the American city.* New York, NY: Russell Sage Foundation.

Massey, Douglas S., & Denton, Nancy A. (1993). *American apartheid.* Cambridge, MA: Harvard University Press.

Meen, Geoffrey, Gibb, Kenneth, Goody, Jennifer, McGrath, Thomas, & Mackinnon, Jane (2005). *Economic segregation in England: Causes, consequences and policy.* Bristol: Policy Press.

Musterd, Sako (2005). Social and ethnic segregation in Europe: Levels, causes, and effects. *Journal of Urban Affairs, 27*(3), 331–348.

Musterd, Sako, & Andersson, Roger (2005). Housing mix, social mix, and social opportunities. *Urban Affairs Review, 40*(6), 761–790.

OECD. (2011). *Divided we stand: Why inequality keeps rising.* Paris: Author.

Peach, Ceri; (with Glazer, Nathan). (1999). London and New York: Contrasts in British and American models of segregation with a comment by Nathan Glazer. *International Journal of Population Geography, 5,* 319–347.

Permentier, Matthieu (2013). Neighbourhood reputations, moving behaviour and neighbourhood dynamics. In M. Van Ham, D. Manley, N. Bailey, L. Simpson, & D. Maclennan (Eds.), *Understanding neighbourhood dynamics: New insights for neighbourhood effects research* (pp. 162–182). Springer: Dordrecht.

Scarpa, Simone (2013). The emergence of a Swedish underclass? Welfare state restructuring, Income inequality and residential segregation in Malmö, 1991–2008. *Economia & Lavoro, (2013)*(2), 121–138.

Van Ham, Maarten, & Clark, William A.V. (2009). Neighbourhood mobility in context: Household moves and changing neighbourhoods in the Netherlands. *Environment and Planning A, 41*(6), 1442–1459.

Wacquant, Loïc (2008). *Urban outcasts: A comparative sociology of advanced marginality.* Cambridge, MA: Polity Press.

Watson, Tara (2009). Inequality and the measurement of residential segregation by income in American neighborhoods. *Review of Income and Wealth Series, 55*(3), 820–844.

Wilson, William Julius. (1987). *The truly disadvantaged: The inner city, the underclass, and public policy.* Chicago: University of Chicago Press.

Appendix

Table A1. Results from logistic regression of the likelihood of moving into targeted neighborhoods (1) versus moving into other neighborhoods in Malmö (0) (1990–1995 and 2005–2010).

Variable	1990–1995			2005–2010		
	B	S.E.	Exp(B)	B	S.E.	Exp(B)
Male	.132	.023	1.141***	.234	.021	1.264***
Female (ref)						
Age over 33 (1990)	−.240	.023	.787***	−.764	.022	.466***
Age under 34 (ref)						
Single with child(ren) 1995	.032	.039	1.032	−.634	.040	.531***
Couple without child(ren) 1995	.022	.039	1.022	−.317	.035	.728***
Couple with child(ren) 1995	.333	.025	1.396***	−.884	.024	.413***
Single without child 1995 (ref)						
New in Malmö 90–95	0.18	.028	1.198***	−.258	.028	.772***
Not new in Malmö LM (ref)						
Born in West Eur	.601	.134	1.824***	1.112	.119	3.042***
Born in East Eur	1.254	.115	3.503***	1.452	.099	4.272***
Born in non-West	1.369	.159	3.933***	1.740	.093	5.699***
Born in Sweden (ref)						
Employed in 1995	−.560	.043	.571***	−.300	.039	.741***
Not employed (ref)						
Low income 1995	.425	.057	1.530***	1.184	.053	3.267***
Middle income 1995	.395	.042	1.483***	.870	.042	2.387***
High income (ref)						
Education 1995 less than 11 ys	.519	.031	1.681***	.005	.027	1.005
Education 1995 11–14 ys	−.015	.029	0.985	−.255	.023	.775***
Education 1995 15 ys+ (ref)						
Interaction West * low inc	.431	.144	1.539***	.569	.127	1.767***
Interaction West * middle inc	.05	.16	1.051	.244	.139	1.277*
Interaction East * low inc	1.389	.12	4.009***	.778	.107	2.177***
Interaction East * middle inc	.08	.132	1.084	.340	.110	1.405***
Interaction non-West * low inc	1.171	.163	3.226***	1.214	.099	3.366***
Interaction non-West * middle inc	.141	.174	1.132	.124	.104	1.132
Constant	−3.123	.062	.044***	−2.708	.057	.067***
Nagelkerke R square	0.3			.36		
−2 log likelihood	62,885			69,309		

* p < 0.05; **p < 0.01; ***p < 0.001.

Table A2. Results from logistic regression of the likelihood of staying in targeted neighborhoods (1) (1990–1995 and 2005–2010) versus moving out from these neighborhoods (0) (1990–1995 and 2005–2010).

Variable	1990–1995			2005–2010		
	B	S.E.	Exp(B)	B	S.E.	Exp(B)
Male	.003	.025	1.003	−.020	.024	.980
Female (ref)						
Age over 33 (1990)	.799	.025	2.224***	.799	.025	2.224***
Age under 34 (ref)						
Single with child(ren) 1995	.461	.04	1.586***	.500	.042	1.648***
Couple without child(ren) 1995	.404	.041	1.497***	.4026	.046	1.495***
Couple with child(ren) 1995	0.117	.028	1.124***	−.022	.028	.978
Single without child 1995 (ref)						
Changed family status 1990–95	−.550	.027	.577***	−.776	.026	.460***
Did not change fam. status (ref)						
Born in West Eur	−.482	.099	.618***	.202	.157	1.224
Born in East Eur	−.196	.084	.822**	.232	.082	1.261***
Born in non-West	−.580	.12	.560***	.233	.089	1.262***
Born in Sweden (ref)						
Employed in 1995	.58	.046	1.785***	−.252	.051	.778***
Not employed (ref)						
Low income 1995	.564	.064	1.757***	.224	.065	1.251***
Middle income 1995	.142	.042	1.152***	.220	.045	1.246***
High income (ref)						
Inc. change 05–10 quartile 1	.335	.045	1.398***	−.027	.043	.973
Inc. change quartile 2	.568	.051	1.765***	.390	.040	1.476***
Inc. change quartile 3	.433	.036	1.542***	.227	.037	1.254***
Inc. change quartile 4 (ref)						
Education 1995 less than 11 ys	.903	.039	2.466***	.499	.033	1.647***
Education 1995 11–14 ys	.616	.037	1.852***	.275	.030	1.317***
Education 1995 15 ys+ (ref)						
Interaction West * low inc	.658	.12	1.930***	−.166	.174	.847
Interaction West * middle inc	.625	.128	1.868***	.325	.191	1.384*
Interaction East * low inc	.556	.095	1.744***	.423	.095	1.527***
Interaction East * middle inc	.357	.101	1.429***	.371	.094	1.449***
Interaction non-West * low inc	1.37	.129	3.935***	.457	.098	1.579***
Interaction non-West * middle inc	.873	.141	2.395***	.413	.100	1.511***
Constant	−1.862	.059	.155***	−.399	.069	.671***
Nagelkerke R square	.186			.21		
−2 log likelihood	41,771			43,429		

*p < 0.05; **p < 0.01; ***p < 0.001.

Are neighbourhoods dynamic or are they slothful? The limited prevalence and extent of change in neighbourhood socio-economic status, and its implications for regeneration policy

Rebecca Tunstall

Centre for Housing Policy, University of York, Heslington, York, UK

ABSTRACT

This article describes what we know about neighbourhood change, and regeneration policy intended to encourage it, using the example of the past 15 years in England. Then it introduces new data on unemployed and middle-class residents as a proportion of all residents in all neighbourhoods in England and Wales over the periods 1985–2005 and 2001–2011. Neighbourhoods are generally slothful rather than dynamic. Thus, we should expect significant change for significant numbers of neighbourhoods only over the long term, and longer time periods than standard for regeneration policy time. This provides important new context for policymaking and evaluation. In this context, we could see the best of neighbourhood regeneration as remarkably successful in creating measureable change against the odds, and as a very valuable part of public policy. Alternatively, we could also see neighbourhood regeneration policy as generally doomed to fail to transform the relative position of neighbourhoods, and as not worth pursuing.

Introduction

Hulchanksi argued that when assessing urban change, 35 years is "not a long time" (Hulchanksi, 2010, p. 7). This article presents evidence that most neighbourhoods are slothful, rather than "dynamic". Meanwhile, "a week is a long time in politics", as the British Prime Minister Harold Wilson (in government 1964–1970, 1974–1976) once said. In practice, in neighbourhood regeneration policy, which aims to reverse negative neighbourhood dynamics and encourage more positive change, few individual projects last as long as a decade. This mismatch between slothful neighbourhoods and dynamic expectations creates dilemmas in policymaking and in evaluation.

First, this article aims to outline some of what we know about neighbourhood change, which provides the context for assessing regeneration efforts and their additionality. Second, it aims to describe policy intended to create neighbourhood change in declined and deprived neighbourhoods, using the example of policy over the past 15 years in England. It sets out the assumptions on which policy has been based, and how it has been evaluated—both praised for creating change and condemned for failing to "transform" neighbourhoods. Third, it aims to provide what is a necessary but until now an absent context for making and evaluating regeneration policy: data on the extent and prevalence of neighbourhood change over time, for all neighbourhoods in a polity, including the majority that have not been subject to regeneration policy. In general, and with or without policy, are neighbourhoods dynamic—or slothful? The paper defines neighbourhood change in relation to population and socio-economics, rather than housing stock value or condition, public space quality, public service quality, business and employment or other indicators. It introduces new data on neighbourhood dynamics across all neighbourhoods in England and Wales, for two economic status variables that are core to understandings of deprivation, gentrification and regeneration: the proportion of residents who are unemployed and the proportion that are in "middle-class" employment. In conclusion, the article suggests that we should expect significant change for significant numbers of neighbourhoods only over time periods longer than standard regeneration policy time periods. Thus, neighbour-hood regeneration policy is generally doomed to fail to transform neighbourhoods' relative socio-economic status. On the other hand, neighbourhood regeneration policy can be seen as remarkably impactful against the odds whenever it creates measureable socio-economic change. In addition to socio-economic change, the focus here, regen-eration policy is also on record as creating measurable change in housing stock value or condition, public space quality, public service quality, business and jobs and other indicators.

Neighbourhood change

A huge amount has been written about neighbourhood change (e.g., see Lupton & Power, 2004 for one review and Megbolugbe, Hoek-Smit, & Linneman, 1996 for another). Scholars have explored types of change, causes of change and the implications for policy towards neighbourhoods in decline, whether in terms of physical conditions, reputation or the social status of residents.

While theoretical approaches have attempted to produce theories that would apply to all neighbourhoods, the focus of empirical research has been on case studies of individual neighbourhoods, such as three in London (Butler & Robson, 2001), or three in Scotland (Robertson, McIntosh, & Smyth, 2010), or on particular cities, such as Adelaide (Badcock & Cloher, 1981), Toronto (Hulchanksi, 2010) and London (Atkinson, 2000). There is also a very substantial literature covering the particular dynamics of gentrification and decline. The literature on gentrification has built up over 50 years, since Glass's original 1964 definition, as a recent reader indicates (Lees, Slater, & Wyly, 2010). Studies have explored definitions, typologisation by phases, gentrification in cities across the world and whether gentrification can be found outside cities (e.g., Phillips, 2009). However, this research is also generally based on case studies. While gentrifying cases generally appear to be in the minority of neighbourhoods in their cities, there is limited evidence on the extent of gentrification as a phenomenon, and the proportion of all neighbourhoods affected. Lees called for more research on the geography of gentrification (2012).

Similarly, there are numerous studies of neighbourhoods in decline, dating from the identification of the "zone in transition" and "twilight zone" early in the history of urban sociology (Grigsby, Baratz, Galster, & Maclennan, 1987; Park & Burgess, 1925), including more specific processes and areas such as those on the "spiral of decline" in social housing estates (e.g., Tunstall & Power, 1995). In the United States and the United Kingdom, urban renewal policy is as old as the concept of gentrification, with pioneering examples in the 1960s, including the Model Cities Program and the Urban Programme, and there is a substantial literature on the efforts to change twilight, declined or deprived neighbourhoods.

Finally, there has been a substantial amount of work describing and typologising neighbourhoods. A private sector industry of neighbourhood typologies has evolved for marketing purposes. However, few typologies incorporate dynamics. Robson et al.'s typology of the 20% most deprived neighbourhoods in England, according to the nature of population flows over one year, was a rare example from the United Kingdom (2009, see also www.ppgis.manchester.ac.uk). Bailey et al. have extended this inquiry with a further typologisation by population dynamics (2013).

As Megubogle et al. said, "The literature is replete with models of neighbourhood change" (1996, p. 1790). These models cover types of trajectories, characteristics and in some cases causes of change. However, relatively little is known about how typical the neighbourhoods that change are, or the prevalence and extent of change across all neighbourhoods. This is an important empirical gap, which means we lack contextual data for planning and assessing neighbourhood policy. This gap may be partly due to the lack of long-run data on neighbourhoods (e.g. Gregory, Dorling, & Southall, 2001; Meen, Nygaard, & Meen, 2013), in turn due to changing administrative and political boundaries in many countries. US and Canadian Census tract boundaries are a significant exception.

The prevalence and rate of neighbourhood change

The extent of the literature on gentrification would lead us to hypothesise that at least a large minority of neighbourhoods might see increasing absolute or relative proportions of middle-class residents, even over a single decade. Critics of regeneration policy have been concerned that whatever the exact theory of change or intent, regeneration may end up creating neighbourhood change through population movement rather than through change for the existing population, and that within the lifetime of projects, what amounts to "state-sponsored" gentrification might occur (e.g. Uitermark & Bosker, 2014). Similarly, the extent of literature on problematic social housing areas might lead us to hypothesise that these areas form at least a large minority if not a majority of social housing areas. However, in a study of the most deprived 30% of neighbourhoods over just one year, when they defined "gentrifier" neighbourhoods as those which had more in-movers from more advantaged areas than others in one year, Robson et al. found these areas made up just 8% of the total (2009). In Toronto, neighbourhood relative position in terms of resident income was largely stable over the short term. Only 9% of neighbourhoods showed consistent increases in income between each of five observations (1980–1990–1995–2000–2005), although 25% showed consistent decreases. A majority of neighbourhoods (60%) experienced change in average incomes relative to the city average of more than 20% only when a period of 35 years was examined (Hulchanksi, 2010). Similarly, survey data show that the vast majority of tenants are satisfied with their homes and neighbourhoods, suggesting spirals of decline affect at most a small group.

On the other hand, another strand of discussion of neighbourhood dynamics in the literature emphasises long-run stability in relative neighbourhood status over change. "Path dependency" or lack of relative change is one of the major causal factors identified in explaining neighbourhood dynamics (e.g. Meen et al., 2013; Robertson et al., 2010). Reputation, levels of disorder and even the moral evaluation of particular neighbourhoods are "sticky" over time (Sampson, 2009, p. 6). Commentary on area regeneration policy in the United Kingdom has noted that there is little change in the rankings of local authorities by relative deprivation (principally measuring resident socio-economic status) over the period in which deprivation indices have been calculated, and despite persistent policy. A number of empirical studies have found evidence of long-run stability in relative neighbourhood status. For example, the relative social status of neighbourhoods in inner London in 1896 correlated highly with measures of deprivation for the same neighbourhoods nearly a century later in 1991, and could be used to predict twentieth-century relative mortality rates (Dorling, Mitchell, Shaw, Orford, & Smith, 2000). In the large majority of travel-to-work-sized areas (much larger than any definition of neighbourhoods), the relative level of infant mortality changed by no more than one quintile from 1891–1900 to 1990–1992, with the same pattern for relative overcrowding and unskilled work (Gregory et al., 2001).

Overall, it is not clear how much neighbourhood change goes on without policy, including in neighbourhoods that have declined and are deprived by some measure. Thus, we don't know how difficult a task of neighbourhood regeneration policy faces when it attempts to instigate change. Information on the prevalence and extent of neighbourhood change would be of importance to the study and theory of neighbourhood dynamics. It would also be of great practical importance, in planning, targeting and evaluating regeneration efforts. It is even of ideological importance, e.g., in assessing the success or failure of social housing or state intervention as a whole.

The practical application of understandings of neighbourhood dynamics: neighbourhood regeneration policy

Understanding of neighbourhood change has direct implications for designing and assessing policy. All industrialised nations have experienced neighbourhood decline and, despite gaps in understanding about the prevalence of decline, its causes and what might reverse it, almost all have seen some policy intended to arrest or reverse this decline. These policies are based on the theory that absolute or relative negative change in physical and/or economic and/or socio-economic attributes of neighbourhoods can be reversed through the application of policy, including public and possibly private funding, by implication, during the lifetime of this policy.

The history—and the recent demise—of neighbourhood regeneration policy in England is presented here as an archetype. As an early industrialising and deindustrialising nation, the United Kingdom has long and extensive experience of neighbourhood decline. At least since the 1960s, and arguably since the Great Depression of the 1930s, there have been a variety of policies attempting to limit or reverse this decline. The most recent comprehensive policy effort was the National Strategy for Neighbourhood Renewal (NSNR) for England. (Neighbourhood renewal and mixed communities policies have always operated slightly differently in Wales, and, to more differently in Scotland and Northern Ireland to policies in England, where the majority of the United Kingdom population live.) The NSNR was launched in 1998 by the then new Labour government, and ran until 2010. The NSNR drew on 30 years of iterative development of policy, and was based on an unusual effort to review existing research and to incorporate researchers and senior practitioners into policymaking (e.g. Cole & Reeve, 2001; Dabinett, 2001). The foundational policy document, *Bringing Britain Together* (Social Exclusion Unit, 1998), combined research evidence with strategy and detailed policy plans. It noted the increase in spatial concentration of poverty and social exclusion in Britain over the 1980s and 1990s. Drawing on research on neighbourhood change, it argued that these problems had structural causes (located outside or at greater spatial scale than the neighbourhood) that mainstream policies had failed these areas, and that past government regeneration efforts had been too small-scale, poorly coordinated and, significantly, too short-term. It pledged that within an extended period of "10–20 years", "no-one should be seriously disadvantaged by where they live" (Social Exclusion Unit, 1998). This pledge did not promise to alter the ranking of any neighbourhood, but did imply that the ranks would be closer together—so close as to make no serious difference.

The NSNR set out 105 commitments, each with measurable indicators, including socio-economic targets such as employment, as well as service outcome measures such as burglary, school results, health, teen pregnancy and jobs and environmental and housing improvements. The strategy combined reliance on macroeconomic growth and public service improvement nationwide, which were to be achieved by "mainstream" policies, with targeted increases in opportunities in deprived areas, labour market changes and area renewal programmes, featuring spending on public space, infrastructure and housing, improved mainstream services and social services such as targeted training. The underlying theory of change was that the interventions would both improve the situation and the quality of life of existing, relatively deprived residents and make it easier for areas to attract less deprived people, resulting in a double absolute and relative impact on area socio-economic mix. Compared to previous urban policy, programmes were longer, lasting up to 10 years.

Deprived areas were to be supported through two major national government funds. First, the Neighbourhood Renewal Fund (NRF) provided funds to the 88 or about one quarter of local authorities with the most deprived populations. Its budget of £0.8bn worked out at under £10 per resident per year, and even if the other public funds drawn into projects are included, spending only totalled about £20 per head per year (author's calculations from AMION, 2010). NRF funds were spent by local authorities and local partners on a wide range of local projects on employment, crime, education and health (AMION, 2010). Second, the New Deal for Communities (NDC) targeted 39 highly deprived neighbourhoods of a few thousand people, all in areas where NRF funding was available. Each carried out over a hundred individual projects on employment, crime, housing, the environment, health and education. NDCs reached less than 1% of the total English population. In the 39 areas where it operated, they spent £1.7bn, or about

£4,000 per head per year (Batty et al., 2010, p. 6). Each of these policies would be expected to reduce local unemployment in absolute—and relative—terms, to help residents gain more middle-class jobs, and to attract more middle-class residents.

Evaluating neighbourhood regeneration policy

The NSNR and NDC and other related programmes operated for 1998–2008/10, a long time in both politics and regeneration policy. These programmes have been evaluated more thoroughly than any previous phases of British urban regeneration. Particular attention has been given to addressing additionality (whether spending produced effects in addition to what might have happened otherwise, particularly given economic growth) and displacement (whether effects in target areas were achieved partly or fully via the movement of less advantaged people out of the areas) (Cole & Reeve, 2001; Dabinett, 2001).

The final evaluation of the NSNR found that differences within local authorities on employment (a socio-economic measure), and education and crime (which could be seen as service outcome measures) reduced over the period 2001–2007 in most areas, even in those without policy interventions (AMION, 2010). However, these gaps reduced more in NSNR local authorities than in similar areas without these programmes (AMION, 2010). This demonstrates additionality and improvement in the relative position for these areas; it does not necessarily imply a change in ranking, however (there could be the same rankings of neighbourhoods, albeit closer together). The final evaluation of the NDCs said that it had "transformed" the few neighbourhoods it operated in (Batty et al., 2010, p. 6). Thirty-two of 36 indicators improved, including those on employment. The gaps between NDC areas, their local authorities and the national average reduced, and NDC areas saw more improvements than other comparator-deprived areas (Batty et al., 2010, p. 6). Again, this demonstrates additionality and relative improvement, although not necessarily change in rankings. Nonetheless, 10 years into the "10–20 years" pledge period, gaps on the 105 targets had not fully closed (AMION, 2010; Batty et al., 2010). In a national study of inequalities in 2010, Hills et al. (2010) recorded persistent substantial gradients by neighbourhood deprivation for socio-economic measures such as income and employment, and service outcome measures, such as health and education.

In summary, the evaluations appear to show that neighbourhood renewal can have measurable and additional impact on neighbourhood socio-economics and on service outcomes. However, this may be smaller in scale than many might have hoped, small relative to continuing inequalities, and does not necessarily result in change in neighbourhood rankings. In a "glass half-full" interpretation, Palmer, MacInnes, and Kenway (2008, p. 19) said:

> ...the successes of the last 10 years need to be stressed in order to confront the damaging idea that everything always gets worse and nothing can be done about it.

However, some commentators argued for a change in direction. For example, by 2010, Hills *et al* argued that neighbourhood renewal "needs renewal" (2010, p. 402).

As it happened, the combination of a new Coalition government in 2010, a budget deficit and the impact of the Global Financial Crisis brought a radical change in

Box 1. Time scales: political, economic.

2 years	Gordon Brown as UK Prime Minister (2008–10)
5 years	Maximum (now fixed) UK Parliamentary Term
10 years	Lifetime of NDCs (1998–2008)
11 years	Tony Blair as UK Prime Minister (1997–2008)
11 years	Economic cycle recession to recession (1981–1992)
10 to 20 years	Period over which NSNR was to have effects
16 years	Economic cycle from end of recession to start of recession (1992–2008)

direction. Despite the adverse effects of the crisis on employment, incomes and investment in disadvantaged neighbourhoods (Hastings, Bramley, Bailey, & Watkins, 2012; Tunstall & Fenton, 2009), the 40-year history of central government-supported neighbourhood regeneration ended abruptly. The NSNR was not extended or replaced. Several bodies associated with neighbourhood regeneration, such as the Regional Development Agencies which assembled land, have been closed. In summary, "neighbourhood renewal... is dead" (Lupton, 2013, p. 66). It could be argued that this change occurred neither despite nor because of the evidence of the results of the policies, but for pragmatic or ideological reasons. However, the absence of more dramatic evidence of large-scale "transformation" within typical policy time spans may be a contributory factor (Box 1).

The context for making and evaluating neighbourhood policy: neighbourhood dynamics in England and Wales, 1985–2011

This paper now introduces evidence of long-term neighbourhood dynamics across all neighbourhoods in England and Wales, which allows us to assess the prevalence and extent of relative neighbourhood change. It aims to explore whether neighbourhoods are dynamic or slothful and to quantify the task policy sets for itself when attempting to instigate change. It compares neighbourhood rankings over time, which allows us to compare the trajectories of different neighbourhoods over different time periods, even where national averages change over time. The time period includes the recession of the early 1990s and the more recent Global Financial Crisis.

Neighbourhood dynamics have been described and measured in terms of house prices and housing condition and other physical and economic characteristics, but there is an established tradition of assessing neighbourhood position and neighbourhood dynamics through population variables (Grigsby et al., 1987). The analysis is based on data for two population variables of interest to those concerned with neighbourhood change and neighbourhood regeneration: (1) unemployment and (2) social class. Employment, unemployment and economic activity rates are important as indicators of the absolute and relative positions of neighbourhoods and of neighbourhood trajectories. They are indirect indicators of the individual and collective income in neighbourhoods, and of levels of real capital and social capital, and material deprivation, and are seen as direct measures of the need for regeneration. For example, in the United Kingdom, labour market measures form one "domain" of the Index of Multiple Deprivation used in England and Wales (and similar indices for Scotland and Northern

Ireland) to target areas for regeneration policy. Absolute and relative proportions of residents of different classes are also important as indicators of the absolute and relative position of neighbourhoods and of neighbourhood trajectories. They are direct indicators of the social status of neighbourhoods, indirect indicators of their individual and collective income, real capital and social capital, and indirect indicators of levels of material deprivation. Second, changes in the class mix of neighbourhoods are a direct indicator of gentrification (Glass, 1964, Smith, 1996; and Lees, Slater, & Wyly, 2007). The class mix of neighbourhoods has been discussed if not formally used as measure of the need for and outcomes of "mixed communities" regeneration policy (e.g. Lupton et al., 2010).

Each of these two variables is examined, pragmatically, for the longest time period with consistent data available. These are 1985–2005 for unemployment and 2001–2011 for social class.

Neighbourhood dynamics in England 1985–2005 in terms of relative rates of unemployment benefit claims

Data and methods

This analysis is based on administrative data on the numbers of people claiming unemployment benefit. "Job-Seeker's Allowance (JSA)" is a benefit available to unemployed working-age adults in the United Kingdom who are available for and "actively seeking" work. It is either means-tested, or paid for by past National Insurance (payroll tax) contributions. Until 1997, the claimant count was an official measure of unemployment. However, eligibility for benefits has been subject to numerous alterations (and, arguably, to political manipulation). Since the late 1990s, data to match the International Labour Organisation (ILO) definition of unemployment has instead been gathered from estimates of unemployment derived from continuous surveys like the Labour Force Survey. From the late 1990s, the claimant count was significantly lower than the ILO measure of unemployment. However, this administrative data is attractive because it is available for repeated observations (monthly or annually), for a long time period—from 1985 to the present, and for small areas.

Fenton created annual working-age population estimates for "postcode sectors" areas for 1985–2001 by matching postal boundaries with small 1981, 1991 and 2001 Census areas (see Tunstall & Fenton, 2009). These population estimates can be used to establish claimant rates for administrative neighbourhoods, called "postcode sectors". There are a total of about 8,000 of these areas in England and Wales, and in 2001 they had an average working-age population of about 4,000. Fenton excluded 896 sectors that had small working-age populations at one or more Census point (fewer than 1,000 people). These were typically city centre commercial zones. Annual population changes between the 1981, 1991 and 2001 censuses were estimated at 10% of the 10-year change. From 2001, he used Office for National Statistics annual Small-Area Population Estimates. There were 6,140 "neighbourhoods" in England and Wales in the final data set, which could be used to create monthly or annual claim rates for small areas. For every year 1985–2005, every postcode sector was assigned to a decile according to the level of unemployment benefit claims relative to the national pattern in that year.

Table 1. Claimant rate decile in 1985 and 2005 for all neighbourhoods in England and Wales, 1985–2005.

						2005 decile					
		1 (Lowest claims)	2	3	4	5	6	7	8	9	10 (Highest claims)
1985 decile	1	44%	26%	17%	6%	5%	1%	1%	0%	0%	0%
	2	25%	24%	22%	12%	8%	5%	2%	0%	0%	0%
	3	16%	21%	21%	17%	12%	6%	4%	2%	0%	0%
	4	9%	16%	18%	21%	16%	11%	7%	3%	1%	0%
	5	2%	6%	13%	21%	21%	16%	12%	7%	2%	0%
	6	2%	3%	7%	13%	19%	22%	18%	11%	5%	0%
	7	0%	2%	3%	7%	12%	21%	22%	21%	11%	3%
	8	0%	1%	0%	2%	4%	14%	24%	27%	22%	6%
	9	0%	1%	0%	1%	2%	4%	9%	22%	35%	26%
	10	0%	0%	0%	1%	0%	1%	2%	6%	26%	66%

Note: Includes only neighbourhoods with unchanged boundaries and full data for the period of 1985–2005, total neighbourhoods = 6,140; rounding means columns and rows may not sum exactly to 100%.

Neighbourhood dynamics in terms of relative rates of unemployment benefit claims

Table 1 shows the trajectories of all neighbourhoods with consistent data in England and Wales in terms of the decile they fitted into in 1985 and in 2005.

Over 20 years of the period 1985–2005, absolute levels of unemployment claims fluctuated markedly. For example, in the decile of areas with the highest rates, they zigzagged from 17% of all adults of working age in 1986, to 9% in 1991, to 15% at a recession peak in 1993 and 4% in 2008 at the low point before the Global Financial Crisis, as the economy changed. However, there was much less change once relative positions were examined. The majority of neighbourhoods were at the same place or a very similar place in 1985 and 2005 in the national ranking (Table 2).

Prevalence, speed and extent of change were all very limited. Relative patterns did not appear to change in the way that absolute ones did. The largest group of neighbourhoods, almost a third of the total (34%), experienced no change in decile over the 10-year period. Twenty per cent of neighbourhoods saw an increase by just one decile, while 18% saw a decrease by just one decile, so a further 38% saw change in either direction of one decile. Thus, a total of 72% of all neighbourhoods saw either no change, or a change of one decile only, insufficient for the neighbourhood to be said to have "transformed" in socio-economic terms, and in all probability not readily

Table 2. Number and percentage of neighbourhoods experiencing different changes in relative proportion of unemployment benefit claimants, England and Wales, 1985–2005.

Change in deciles	Percentage of neighbourhoods
+5	*
+4	1%
+3	4%
+2	8%
+1	20%
0	34%
−1	18%
−2	9%
−3	4%
−4	2%
−5	*

Note: * = less than 1%.

noticeable on the ground, particularly given the marked fluctuations in absolute claimant rates over this time. If concentrations of unemployment create knock-on disadvantage or are symptoms of it, it appears that 10–20 years is not enough to see a serious reduction of this disadvantage.

There was particularly little change for areas that started with the highest unemployment claims. Of the 614 postcode sectors in England and Wales that were in the highest decile for claims in 1985, 400 (65%) were in the top decile again in 2005 (although they may have experienced variation between these dates). This represents an "escape rate" from this group of 1.7% per year. Only eight (1.9%) had below average claim rates in 2005. This represents an "arrival rate" in this half of the distribution of 0.06% per year.

Of the 614 postcode sectors in England and Wales that were in the highest claims decile in 1985, the mean decile throughout the entire period of 1985–2005 was 1.3. A total of 259 (42%) were in the top decile for every year over the period 1985–2005. A total of 257 more were only in the top two deciles. Only 98 (16%) were ever in the third decile or above. Only 13 (2%) had experienced below average JSA rates at any time to 2005.

Even more dramatically, there was also very limited change for areas with high unemployment over a period of 70 years. Again, relative patterns did not appear to change in the way that absolute ones did. We examined local authority (not neighbourhood) areas that had high unemployment in 1934, and which had been identified as "Depressed Areas" in the Special Areas (Development and Improvement) Act 1934. This commenced a form of regional policy, which could be seen as a forerunner of the regeneration policy which began in the 1960s. Of the 251 contemporary neighbourhoods identifiable as located in "Depressed Area" local authorities, only 27 (11%) were in the top decile for unemployment claims in 2005. However, this only means an "escape rate" from this group of 1.3% per year. Only 49 (19%) had below average JSA rates, in other words, an "arrival rate" in this half of the distribution of 0.3% per year.

Neighbourhood dynamics in England (2001–2011) in terms of relative proportions of "middle-class" residents

Data and methods

This analysis was based on data in the 2001 and 2011 Censuses of population for England, the first to offer small area data for boundaries consistent over time. Small neighbourhoods were ranked according to the proportion of "middle-class" residents they had in 2001 and in 2011. Small neighbourhoods were defined as Lower-level Super Output Areas (LSOA), with an average population of 1,500 (smaller than postcode sectors). The analysis excluded neighbourhoods whose boundaries changed over the period 2001–2011. Although as areas with population size change, they might be interesting, they formed just 2.5% of the total. "Middle-class" residents were defined as those aged 16–74 whose current or most recent employment was in categories 1 and 2 according to the National Social and Economics categorisation (for 2001, Large employers and higher managerial occupations/Higher professional occupations and Lower managerial and professional occupations, and for 2011, Higher Managerial, Administrative and Professional Occupations and Lower

Managerial, Administrative and Professional Occupations). This definition has been used in other studies of gentrification (e.g. Davidson & Lees, 2010). For both 2001 and 2011, the ranking of neighbourhoods was divided into 10 deciles, with decile 1 having the lowest proportion of middle-class residents and decile 10 the highest.

Neighbourhood dynamics in terms of relative proportions of "middle-class" residents

The table below shows the 2001 and 2011 deciles for individual neighbourhoods in England and how they compared to each other (Table 3).

Again, as in the case of the employment measure (which related to a longer time period but larger neighbourhoods), the majority of neighbourhoods were at the same place or a very similar place in 2001 and 2011 in the national ranking, according to the proportion of middle-class residents. Again, relative patterns did not appear to change in the way that absolute ones did.

The table below shows the number of neighbourhoods experiencing different extents of changes over the period 2001–2011 according to the relative proportion of "middle-class" residents (Table 4).

The largest group of neighbourhoods, almost half of the total (47%), experienced no change in decile over the 10-year period. Although there was on average a slight

Table 3. Percentage of neighbourhoods experiencing particular trajectories in terms of class, England, 2001–2011.

		2011 decile									
		1 (Least middle class)	2	3	4	5	6	7	8	9	10 (Most middle class)
2001 decile	1	77%	20%	2%	1%	0%	0%	0%	0%	0%	0%
	2	19%	52%	23%	5%	1%	0%	0%	0%	0%	0%
	3	3%	20%	42%	25%	8%	2%	1%	0%	0%	0%
	4	1%	6%	21%	35%	25%	9%	2%	1%	0%	0%
	5	0%	2%	8%	22%	34%	24%	8%	2%	0%	0%
	6	0%	0%	3%	8%	21%	33%	25%	8%	1%	0%
	7	0%	0%	1%	3%	9%	22%	34%	25%	6%	0%
	8	0%	0%	0%	1%	2%	9%	22%	36%	27%	3%
	9	0%	0%	0%	0%	0%	2%	7%	23%	45%	22%
	10	0%	0%	0%	0%	0%	0%	1%	5%	19%	74%

Note: Includes only neighbourhoods with unchanged boundaries 2001–2011; total of 31,675 neighbourhoods; rounding means columns and rows may not sum exactly to 100%.

Table 4. Percentage of neighbourhoods experiencing relative change in the proportion of "middle-class" residents, England, 2001–2011.

Change in deciles	Percentage of neighbourhoods
+4	*
+3	1%
+2	5%
+1	22%
0	47%
−1	19%
−2	6%
−3	1%
−4	*

Note: * = less than 1%.

increase in the proportion of middle-class residents in these areas, it only reflected national trends. A further 22% of neighbourhoods saw an increase by just one decile, and 19% saw a decrease by just one decile. Thus, a further 41% saw change in either direction of one decile. Thus, a total of 88% of all neighbourhoods saw either no change, or a change of one decile only. This change might not be sufficient to be directly or indirectly noticeable on the ground over a 10-year period. Only a small minority of neighbourhoods experienced more "marked" change.

Only 6% of neighbourhoods experienced an increase in the proportion of middle-class residents, enough to change their position by two deciles or more. These might be described as "gentrifying" neighbourhoods. However, it could be argued that the "classic" conception of a gentrified neighbourhood is one in which the "gentry" initially formed a below average proportion of the population but have come to form a proportion above average. If we define it in this way, the total number of neighbourhoods in England experiencing "classic gentrification" between 2001 and 2011 was 791, or 2% of the total. Thus "gentrification", while occupying a very important place in urban studies literature, is an unusual process, and gentrifying neighbourhoods are an atypical and very small minority. Similarly, only 6% of neighbourhoods experienced decline, no change or limited increases (against the national trends of an increasing middle class) in the proportion of middle-class residents enough to change their position by two deciles or more. These might be described as the "declining" neighbourhoods.

Over 10 years of the period 2001–2011, there was little change for areas with very low proportions of middle-class residents. Of the 3,168 LSOAs in England that were in the lowest decile for middle-class residents in 2001, 2,447 (or 77%) were in the lowest decile again in 2011 (although they may have experienced variation between these dates). This represents an "escape rate" from this group of 2.3% per year (a faster rate than for the employment variable). Only four (0.1%) had above average rates of middle-class residents in 2011. This represents an "arrival rate" in this half of the distribution of 0.01% per year (a slower rate than for the employment variable). Of the 3,168 LSOAs in England that were in the lowest decile for middle-class residents in 2001, 3,075 or 97% were only in the top two deciles, and only 3% were ever in the third decile or above. Thus neighbourhoods appeared to be largely "slothful", even in a period that included the Global Financial Crisis.

Setting regeneration policy in the national context of "slothful" neighbourhoods: the case of change in unemployment benefit claimant rates for New Deal Communities areas

Change in claimant rates for the NDC areas provides an example of the potential contribution of data on neighbourhood dynamics. A postcode sector (the area used in the analysis above) is about the same population size as a NDC area, but slightly smaller. For 36 of the 39 NDC areas, a single postcode sector proxy could be confidently identified.

Looking at the pre-policy context, 26 (72%) NDC areas were in the top decile for claimant rates in 1998, when the NDC began. Thus, as intended, the NDC was generally targeting a handful of the hundreds of neighbourhoods in the highest claimant rate

Table 5. Job seekers' allowance claimant rate for NDC areas (NDC operated from 1998 to 2008).

	1985	1998	2005	Every year over the period 1985–1997 (before policy)	Every year over the period 1998–2005 (during policy)
Highest rate decile	24	26	26	18	24
Other	12	10	10	18	12

Note: Data available for 36 of 39 NDC areas.

decile. However, generally, NDC was targeting areas that had experienced not only high unemployment just at one point but sustained high claim rates over time (Table 5).

Twenty-four (66%) NDC postcode sectors had been in the top decile for JSA claims in 1985. There was some variation in where in the top decile these areas were, and 33% of NDC postcode sectors were not in the top decile, so despite being a small group, NDC areas were not uniformly very extreme. Eighteen (50%) had been in the top decile every year in the 12-year period from 1985 to 1997. Thirty-one (86%) were in only the top two deciles over the period 1985 to 1997. Only five had experienced third decile or better at least once over the period 1985–1997.

Did neighbourhoods' relative JSA claim rates change during the NDC policy? No. All those NDC postcode sectors that had been continuously in the top decile before the scheme (1985–1997) were so during and after 1998–2005 too. Twenty-four NDCs were in the top decile for every year from 1998 to 2005. The mean position for all NDCs form 1998 to 2005 was the same as that from 1985 to 1997. Twenty-eight NDC postcode sectors were in the top decile for JSA claims in 2005, when the NDC was in its final years.

Considered as additional evaluative material, this data on long-term neighbourhood trajectories nationwide provides a negative assessment of NDCs' ability to change neighbourhoods' ranking. It perhaps provides a more negative assessment than the far more complex evaluations that recognised multiple aims and drew on multiple indicators and sources of evidence (e.g. Batty et al., 2010). However, considered as contextual material, this data on long-term neighbourhood trajectories nationwide demonstrates how rare "escapes" from prolonged relatively high claimant rates (or low middle-class population rates) are. This emphasises the difficulty of the task being set for regeneration policy, and the value of the limited changes produced. If the limited change in NDC areas appears disappointing, it should be noted that of the 241 non-NDC areas continuously in the top decile 1985–1997, not a single one "escaped" even for one year during the NDC period 1998–2005. Thus, a change of this extent (a whole decile) appeared impossible or very unlikely without policy, and policy was setting itself a great challenge in attempting to achieve this kind of transformative change (Box 2).

The implications of evidence on neighbourhood change

This evidence supports the "slothful" school of research above the "dynamic neighbourhoods" school. More research would be valuable on the causes of lack of change, and determination of which neighbourhoods change and which do not, to complement the extensive literature on causes of change and descriptions of path dependency. We need to know which neighbourhoods break from the path and which do not. In addition,

Box 2. Time scales: neighbourhood change, with and without policy.	
7 years	Of the 24 NDC areas continuously in the top decile for claimant rates in the decade before NDC policy started (1985–97), not a single one "escaped" during the NDC period (1998–2005), even for one year
	Of the 241 non-NDC areas continuously in the top decile for claimant rates in the decade before NDC policy started (1985–1997), not a single one "escaped" during the NDC period (1998–2005), even for one year
10 years	Gaps between neighbourhoods on wide range of indicators, including employment, reduced more in NSNR local authorities than in similar areas without these programmes (AMION, 2010). Gaps between NDC areas, their local authorities and the national average on wide range of indicators, including employment, reduced and NDC areas saw more improvements than other comparator-deprived areas (Batty et al., 2010, p. 6)
	Only 0.1% of small neighbourhoods in England and Wales in the decile with the lowest proportion of middle-class residents achieved average levels
20 years	Two per cent of neighbourhoods in England and Wales in the decile with highest unemployment benefit claims experienced average rates
35 years	Majority of neighbourhoods in Toronto experienced change in average income of 20% or more of city average; 9% experienced continuous improvement in relative position over five observations (Hulchanksi, 2010)
70 years	Nineteen per cent of small neighbourhoods in 1934 "Depressed Areas" in England and Wales achieved below average unemployment benefit claim levels
c90–100 years	Minority of larger areas changed by more than one quintile in relative status on infant mortality, overcrowding and low-skilled employment (Gregory et al., 2001)

there are implications for policy design and evaluation. These long-term data on neighbourhood trajectories can be used to provide important additional evaluation and context material for neighbourhood regeneration policy. Both sets of analysis above confirm the arguments and limited empirical evidence from the literature that suggests long-term stability in neighbourhood relative status is the norm, and that marked change in neighbourhood rankings is rare. The data extends arguments and evidence in the literature by quantifying the prevalence and extent of change for two important variables.

Given the low "escape rates" from high relative unemployment and low relative middle-class populations, even through periods of recession and Global Financial Crisis, and given the size and scope of typical regeneration projects, significant change in relative socio-economic position can only be realistically predicted for the most deprived areas over a period of multiple decades, which is very long by the standards of politics and regeneration policy to date.

Significant change over significant numbers of neighbourhoods takes place over time periods longer than standard political or policy time. This might have provided alternative viewpoints on NSNR for commentators and for policymakers of all parties. In summary, we should not expect many neighbourhoods to "transform" in relative socio-economic status spontaneously in the short and medium terms. We should not expect policy aimed at the most disadvantaged neighbourhoods to "transform" them in socio-economic terms a similar time period, although improvements to housing conditions, services and public spaces may be achievable. Thus, perhaps both Labour and Coalition policymakers, who became frustrated with the speed of results from neighbourhood renewal policy in England over the 2000s, were being too hasty in their judgements.

However, there is always an additional step of judgement to be taken between research evidence and policy. There are two possible and contradictory policy implications to be drawn from this data. On the one hand, we could see the best of neighbourhood regeneration as remarkably successful in creating measureable change against the odds, and thus as a very valuable part of public policy. On the other hand, we could also see neighbourhood regeneration policy as generally doomed to fail to transform the relative socio-economic position of neighbourhoods, and thus not worth pursuing. Batty et al. (2010) pointed out there was scope for more energetic regeneration policy, which might have more transformative results. Even the well-funded and exceptional NDCs represented an addition of just 10% on general public expenditure in its areas. In fact, the United Kingdom is currently undergoing a natural experiment in the opposite approach: a "policy off" period with minimal regeneration policy.

Disclosure statement

No potential conflict of interest was reported by the author.

Funding

The work on the numbers of people claiming unemployment benefit draws heavily on work carried out by and with Alex Fenton of LSE, supported by the Joseph Rowntree Foundation, and partly published as Tunstall and Fenton (2009).

References

AMION. (2010). *Evaluation of the national strategy for neighbourhood renewal: Final report.* London: Communities and Local Government.

Atkinson, R. (2000). Measuring gentrification and displacement in Greater London. *Urban Studies, 37*, 149–165.

Badcock, B.A., & Cloher, D. Urlich (1981). Neighbourhood change in Inner Adelaide, 1966-76. *Urban Studies, 18*(1), 41–55.

Bailey, Nick, Barnes, Helen, Livingston, Mark, & Mclennan, David (2013). Understanding neighbourhood population dynamics for neighbourhood effects research: A review of recent evidence and data source developments. In Maarten van Ham, David Manley, Nick Bailey, Ludi Simpson, & Duncan Maclennan (Eds.), *Understanding neighbourhood dynamics* (pp. 13–41). Dordrecht: Springer.

Batty, Elaine, Beatty, Christine, Foden, Mike, Lawless, Paul, Pearson, Sarah, & Wilson, Ian (2010). *The new deal for communities experience: A final assessment the new deal for communities evaluation: Final report – volume 7.* London: Communities and Local Government.

Butler, Tim, & Robson, Gary (2001). Social capital, gentrification and neighbourhood change in London: A comparison of three South London neighbourhoods. *Urban Studies, 38*, 2145–2162.

Cole, Ian, & Reeve, Kesia (2001). *New deal for communities: National evaluation scoping phase: Housing and physical environment domain: A review of the evidence base.* Sheffield: Sheffield Hallam University.

Dabinett, Gordon (2001). *A review of the evidence base for regeneration policy and practice.* London, UK: DETR.

Davidson, Mark, & Lees, Loretta (2010). New-build gentrification: Its histories, trajectories, and critical geographies. *Population, Space and Place, 16*, 395–411.

Dorling, Danny, Mitchell, Richard, Shaw, Mary, Orford, Scott, & Smith, George Davey (2000). The ghost of Christmas past: health effects of poverty in London in 1896 and 1991. *BMJ, 321* (7276), 1547–1551.

Glass, Ruth (1964). Introduction: Aspects of change. In Centre for Urban Studies (Ed.), *London: Aspects of change* (pp. xiii–xlii). London: McGibbon and Kee.

Gregory, Ian N., Dorling, Danny, & Southall, Humphrey R. (2001). A century of inequality in England and Wales using standardized geographical units. *Area*, 33(3), 297–311.

Grigsby, William, Baratz, Morton, Galster, George, & Maclennan, David (1987). *The dynamics of neighbourhood change and decline.* Oxford: Pergamon.

Hastings, Annette, Bramley, Glen, Bailey, Nick, & Watkins, David (2012). *Serving deprived communities in a recession.* York: JRF.

Hills, John, Brewer, Mike, Jenkins, Stephen, Lister, Ruth, Lupton, Ruth, Machin, Steve, ... Riddell, Sheila (2010). *An anatomy of economic inequality in the UK: The report of the National Equality Panel.* London: Government Equalities Office, Communities and Local Government.

Hulchanksi, J. David (2010). *The three cities within Toronto: Income polarization among Toronto's neighbourhoods, 1970–2005.* Toronto, ON: Cities Centre, University of Toronto.

Lees, Loretta. (2012). The geography of gentrification: Thinking through comparative urbanism. *Progress in Human Geography*, 36, 155–171.

Lees, Loretta, Slater, Tom, & Wyly, Elvin (2007). *Gentrification.* London: Routledge.

Lees, Loretta, Slater, Tom, & Wyly, Elvin (2010). *The gentrification reader.* London: Routledge.

Lupton, Ruth (2013). What is neighbourhood renewal policy for? *People, Place & Policy Online*, 7 (2), 66–72.

Lupton, Ruth, Heath, Natalie, Fenton, Alex, Clarke, Anna, Whitehead, Christine, Monk, Sarah, ... Robinson, Jamie (2010). *evaluation of the mixed communities initiative demonstration projects.* London: Communities and Local Government.

Lupton, Ruth, & Power, Anne (2004). *What we know about neighbourhood change: A literature review.* CASE Report 27. London, UK: Centre for the Analysis of Social Exclusion, London School of Economics and Political Science.

Meen, Geoffrey, Nygaard, Christian, & Meen, Julia (2013). The causes of long-term neighbourhood change. In Maarten Van Ham, David Manley, Nick Bailey, Ludi Simpson, & Duncan MacLennan (Eds.), *Understanding neighbourhood dynamics: New insights for neighbourhood effects research* (pp. 43–52). Springer: Dordrecht.

Megbolugbe, Isaac F., Hoek-Smit, Marja C., & Linneman, Peter D. (1996). Understanding neighbourhood dynamics: A review of the contributions of William G. Grigsby. *Urban Studies*, 13, 1779–1795.

Palmer, Guy, MacInnes, Tom, & Kenway, Peter (2008). *Housing and neighbourhoods monitor 2008.* York: JRF.

Park, Robert E., & Burgess, Ernest W. (1925). *The city.* Chicago, IL: University of Chicago Press.

Phillips, M. (2009). Counterurbanisation and rural gentrification: An exploration of the terms. *Population, Space and Place*, 16(6), 539–558.

Robertson, Douglas, McIntosh, Ian, & Smyth, James (2010). Neighbourhood identity: The path dependency of class and place. *Housing, Theory and Society*, 27(3), 258–273.

Robson, Bryan, Lymperopoulou, Kitty, & Rae, Alaisdair (2009). *A typology of the functional roles of deprived neighbourhoods.* London: Department of Communities and Local Government.

Sampson, Robert J. (2009). Disparity and diversity in the contemporary city: Social (dis)order revisited. *British Journal of Sociology*, 60(1), 1–31.

Smith, Neil (1996). *The new urban frontier: Gentrification and the revanchist city.* New York, NY: Routledge.

Social Exclusion Unit. (1998). *Bringing Britain together: A national strategy for neighbourhood renewal.* London: HMSO.

Tunstall, Rebecca, & Fenton, Alex (2009). *Communities in recession.* York: JRF.

Tunstall, Rebecca, & Power, Anne (1995). *Swimming against the tide?* York: JRF.

Uitermark, Justus, & Bosker, Tjerk (2014). Wither the 'undivided city'? An assessment of state-sponsored gentrification in Amsterdam. *Journal of Economic and Social Geography*, 105(2), 221–230.

Index

www.ingramcontent.com/pod-product-compliance
Ingram Content Group UK Ltd.
Pitfield, Milton Keynes, MK11 3LW, UK
UKHW010020280225
455677UK00023B/719